GENE TARGETING

GENE TARGETING

Edited by

MANUEL A. VEGA

CRC Press
Boca Raton Ann Arbor London Tokyo

Library of Congress Cataloging-in-Publication Data

Gene targeting / edited by Manuel A. Vega.
p. cm.
Includes bibliographical references and index.
ISBN 0-8493-8950-X
1. Genetic engineering. I. Vega, Manuel A.
QH442.G4384 1994
575.1′0724—dc20 94-14299
CIP

Direct all inquiries to CRC Press, Inc., 2000 Corporate Blvd., N.W., Boca Raton, Florida 33431.

International Standard Book Number 0-8493-8950-X
Library of Congress Card Number 94-14299
Printed in the United States of America 1 2 3 4 5 6 7 8 9 0
Printed on acid-free paper

Dedication

To

Lila, Manuel, Ana, Francisco,

Teresa, and Juan

Acknowledgments

The Editor specially thanks Drs. Claude Besmond and Michel Goossens (U.91-INSERM, Paris), the AFLM (Association Française de Lutte contre la Mucoviscidose), Professor Stuart G. Siddell (Wurzburg University, Germany), and SECECOM (Computer Service Center—CONICET), Bahia Blanca, Argentina.

The Editor

Manuel A. Vega, Ph.D., is chief of the Laboratory of Gene Therapy, Hospital Interzonal Dr. J. Penna, Bahia Blanca; Adjunct Professor of Medical Molecular Genetics, Universidad Nacional del Sur (UNS), Bahia Blanca; and Researcher of the National Research Council, Argentina.

Dr. Vega graduated in 1982 from the Department of Biology, UNS, and obtained his Ph.D. degree in 1987 from UNS. He did postdoctoral research at the Institut fur Virologie and Immunobiologie, Wurzburg University, Germany; at the Laboratory de Genetique Moleculaire, U.91-INSERM; Hospital H. Mondor, Creteil (Paris), France; and at the Department of Gene Therapy, TNO, Rijswijk, The Netherlands; as Wissenschaftlicher Mitarbeiter, fellow of AFLM (French Association against Cystic Fibrosis), and fellow of EMBO (European Molecular Biology Organization), respectively.

Among other awards, he has received the Dr. E. DeRobertis award from the State Secretariat for Science and Technology, Argentina, and the Dr. J.A. Balseiro award from the Forum for Science and Technology for Production from the State Ministry of Culture and Education, Argentina. He has been a recipient of research grants from the National Research Council, the Smith-Klein Beecham Foundation, the A.J. Roemmers Foundation, UNS, and private industry.

Dr. Vega has presented over 20 invited lectures at international meetings, institutes, and universities and he has published 16 research papers. His current major research interests include gene therapy for cystic fibrosis and cancer and prevention of cystic fibrosis by gamete selection.

The Contributors

Wolf M. Bertling
Paul-Ehrlich Institute
Paul-Ehrlich-Strasse
Langen, Germany

Anegela K. Cruz
Faculdade de Odontologia de Ribeira Preto
Faculdade de Medicina de Ribeira Preto
Universidade de São Paulo
São Paulo, Brazil

Jean-Louis Guenet
Unite de Genetique des Mammiferes
Institut Pasteur de Paris
Paris, France

Paul Hooykaas
Institute of Molecular Plant Sciences
Leiden University
Clusius Laboratory
Leiden, The Netherlands

Stephen C. Kowalczykowski
Division of Biological Sciences
Sections of Microbiology and of
Molecular and Cellular Biology
University of California, Davis
Davis, California

Thomas Lufkin
Brookdale Center for Molecular Biology
The Mount Sinai Medical Center
New York, New York

Remko Offringa
Institute of Molecular Plant Sciences
Leiden University
Clusius Laboratory
Leiden, The Netherlands

Manuel A. Vega
Laboratory of Gene Therapy
Department of Biology
Universidad Nacional del Sur
Bahia Blanca, Argentina

Alan S. Waldman
Department of Biological Sciences
University of South Carolina
Columbia, South Carolina

David A. Zarling
Cell and Molecular Biology Laboratory
SRI International
Menlo Park, California
Department of Laboratory Medicine
University of California, San Francisco
San Francisco, California

Table of Contents

Chapter 1

Gene Targeting

Wolf M. Bertling

Paul-Ehrlich Institute

Paul-Ehrlich-Strasse.

D-6070 Langen, Germany

Contents

0-8493-8950-X/95/$0.00+$.50

I. Introduction

For some time now, genome manipulation of pro- and eukaryotes and the technologies involved have been the focus of study in many labs. The expression and analysis of foreign DNA in prokaryotes is the most widely known application of these technologies. However, since these techniques have been improved and adjusted, they now allow for the introduction of such minor changes as point mutations in genomes as complex as the human genome. Especially in the field of research of higher eukaryotes, there have been many attempts to use our newly acquired knowledge to produce drugs in manipulated cells, to generate new models for diseases, and to, ultimately, perform somatic gene therapy.

Here we will give an introduction to recombination in general and to homologous recombination in particular. For this purpose, we will define the basic terminology and point out the importance of gene targeting in the mammalian genome. A brief inspection of the historical developments of this genetic tool, particularly of the problems that are associated with it, will lead to the description of current recombination models. This will be followed by new thoughts on the occurrence of homologous recombination events *in vivo* and their mechanisms, such as the involvement of repetitive elements and RNA in homologous recombination. We will discuss in detail the parameters of frequency and fidelity and will explain possible approaches to work around the inherent inefficiency of the process causing gene targeting. Finally, we will demonstrate the use of gene targeting for a wide variety of genes and purposes. Only seemingly of secondary importance are the different mechanisms for transporting DNA from the outside to the inside of a cell and into the nucleus. Therefore, a brief listing of current technologies to introduce DNA into the nuclei of cells in gene targeting experiments will also be presented. Since gene targeting in bacteria, lower eukaryotes, and plants will be discussed extensively in other chapters of this book, I will refer mainly to the development of gene targeting in mammalian cells.

II. Terminology

The term gene targeting refers to a targeted alteration of a specific DNA sequence in its genomic locus and occurs as a result of the homologous recombination of chromosomal and extrachromosomal sequences. Recombination, the exchange of genetic information, is the basic element of gene targeting, previously also defined as reassortment of a series of nucleotides along nucleic acid molecules.[1] Whereas in a non-homologous or illegitimate recombination, genetic elements of no significant homology or similarity recombine, homologous recombination is a process in which two DNA entities with a high sequence homology interact and recombine, i.e., exchange genetic information by adding or replacing their sequence elements. Generally, gene targeting is considered the recombinational interaction of an exogenous extrachromosomal and an endogenous chromosomal sequence by means of human manipulation. This is in contrast to the mainly naturally occurring processes of sister chromatide exchange or intrachromosomal recombination and is also different from the extrachromosomal recombination between two mostly exogenous, extrachromosomal elements. Every one of these processes can lead to either insertion or replacement of homologous sequences. Several reviews on gene targeting and homologous recombination, emphasizing different aspects, have been published in recent years.[2-13]

According to a definition of Koller and Smithies,[12] there are three possible mechanisms of a gene targeting reaction. For the shape of these molecules, specifically the mechanism, Smithies names these processes O- and Ω-type events, and two-step or in-and-out mechanism (Figure 1). Genetic information can be targeted with a single cross-over event, as first described by Hinnen et al.[14] and more extensively studied by Orr-Weaver et al.,[15] leading to an insertion of the new piece of DNA into the existing sequence. A replacement due to a double cross-over was first described by Rothstein.[16] If such a replacement construct is shorter, its integration creates a deletion, or if it is longer, it creates an insertion. If two excision cross-over events occur successively, the recombination product may have only a minute difference to the original.

Similar to the replacement process is the so-called gene conversion, which leads to the adaptation of the sequence of one strand to the sequence of another strand. Gene conversion is generally defined as locally restricted copying of genetic information from one strand to another. Therefore, it may be considered as a different form of homologous recombination, not necessarily being associated with cross-overs and, therefore, being non-reciprocal (Figure 2). The main difference is that homologous recombination products resulting from one or more cross-over events are reciprocal, because both parental double strands carry elements from another after the recombination. These two concepts of recombination are frequently associated, since both involve DNA breakage and repair and the enzymes necessary for these events.[17] Enzymatic activities involved in DNA breakage and repair were known before to occur in

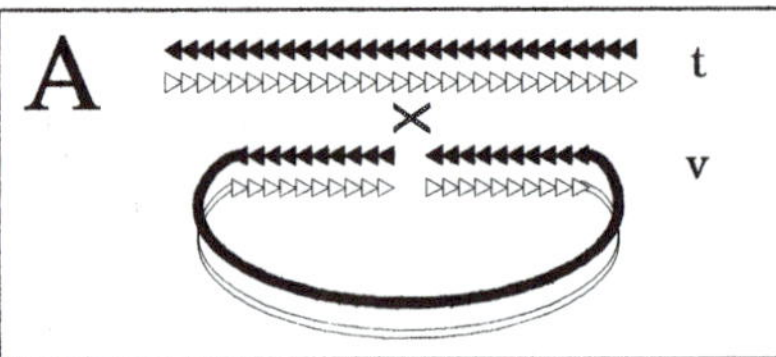

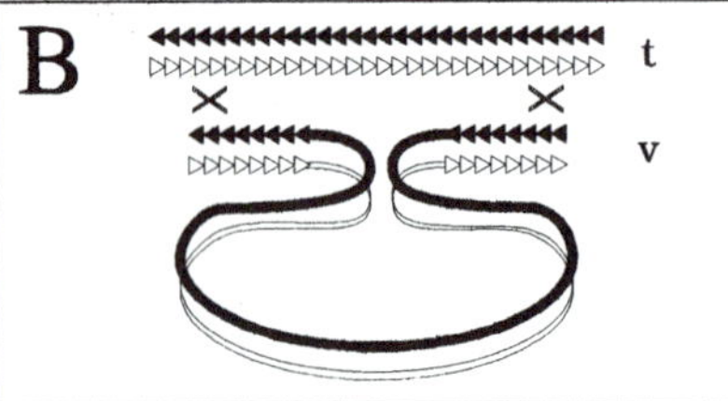

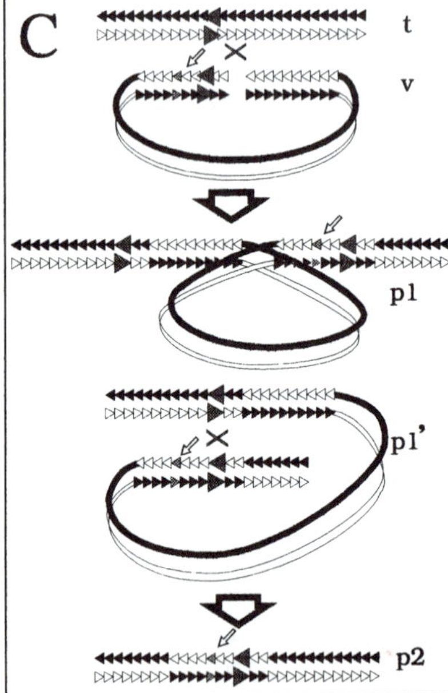

Figure 1. Schematic drawing of three forms of gene targeting. ***A*** *shows a single-step O-type insertion event. This type of insertion leads to a duplication of the target sequences. The chromosomal DNA ((t), upper two lanes of triangles) inserts the construct (v), which is linearized in the homologous part. The non-homologous parts are shown as black and white lines, respectively. The Ω-type replacement in* ***B*** *involves two sites of recombination. Note that no duplication of the target site results and that the orientation of the marker (in the non-homologous loop) is opposite to the product one would receive in* ***A****. In the so-called in-out targeting procedure, two successive steps are necessary* ***(C)****. After an initial O-type recombination, a primary insertion product forms* ***(p1)****. The large dotted triangles symbolize specific marker areas; the small dotted triangles symbolize minute sequence differences in the homologous region. The genetic marker is duplicated after the first cross-over. Note that the duplicated sequences are inverted compared to the origional. After reshaping of this product* ***(p1)****, a new recombination leads to the desired product* ***(p2)*** *by excision cross-over. It resembles the original sequence, except for the mutation, e.g., a point mutation (small dotted triangles).*

human cells.[18,19] The phenotypical and mechanistic similarities of gene conversion and homologous recombination will be discussed in greater detail when the current models are introduced.

Different from these processes is another class of events: site specific recombination (for review, see Reference 20). Although a targeted DNA segment also recombines at a specific site, which in this case shows an extensive nucleotide sequence homology, it especially involves mechanisms that target sequences to either short specific signal sequences or even to sites where specific proteins are present. Sequence pattern recognition is used for the integration of λ in *E. coli*, as well as for the insertion of other viruses into pro- and eukaryotic genomes, for the rearrangement of vertebrate immunoglobulines, and to a certain extent, for the movement of transposable elements. One group of transposable elements, the so-called retrotransposons, however, seems to target either sites similar to the ones that they originated from or sites where specific proteins such as RNA polymerases or DNA polymerases and related factors bind to DNA.

Due to the low complexity of bacterial genomes, many basic mechanisms and factors involved in recombination have been identified and studied in bacteria, especially *E. coli,*[2] where the majority of these proteins has been characterized. Much of the information currently available about the genetics of eukaryotic recombination has been generated in lower eukaryotes such as yeast and other low fungi. This information is now used to explore new ways to allow for the genetic manipulation of mammals, an admittedly anthropocentric goal.

III. Historical developments and current recombination models

Relatively early in the 1900s, recombination was seen as a means to generate species diversity, although the elements to comprise the genetic information were unknown.

In the 1960s and 1970s, other terms for homologous recombination had been introduced, such as general, generalized, chromosomal or equatorial.[21] Some of these terms, such as normal or non-specific,[22-25] would be considered misleading nowadays, considering that, in gene targeting attempts, the majority of constructs insert at random sites[7] rather than at the predetermined, homologous site. So it is not normal to detect a homologous recombination event in which an exogenous construct integrated at a specific site, the homolo-

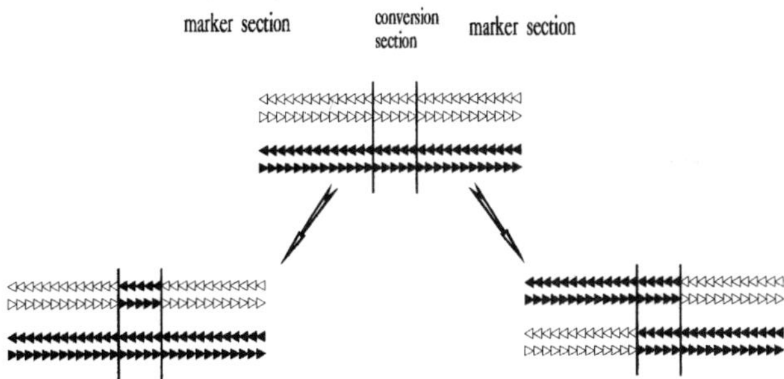

Figure 2. *The principle of gene conversion events with or without accompanying cross-over. Marker sections of two homologous "white" (target) and "black" double-stranded DNAs are separated by vertical lines from the section of conversion. The non-reciprocal product on the lower left was not subject to a cross-over, whereas a reciprocal exchange accompanied the conversion event of the product on the lower right.*

gous region. However, the situation is different *in vivo,* and interchromosomal recombination, for instance, is primarily homologous.

In 1928, Griffith[26] observed genetic recombination in bacteria and termed it transformation. In the early 1960s, the selection process of a homologous recombination reaction was first attributed to basepairing of complementary single strands to generate a heteroduplex region near the point of exchange.[27,28] This led to the development of molecular models for the mechanism underlying recombination, which is discussed in detail in Chapter 2.

A. Conventional models

The oldest and most significant model to describe homologous recombination is the Holliday model (Figure 3). A so-called Holliday structure, formed as an intermediate of the recombination process, resolves either with or without cross-over events. Fluorescence energy transfer studies indicated that the four-way DNA junction in such a Holliday structure is a right-handed cross of antiparallel molecules.[29]

This model uses the same basic structure to explain gene conversion and cross-over-driven homologous recombination. The heteroduplex in the case of gene conversion has to be subjected to repair enzymes to clear mismatched bases. So both concepts ask for some cutting and religation activities acting on the DNA strands. If such a heteroduplex is transient and moves along the DNA, which may resolve with or without a preceding cross-over, we are following Sobell's model.[30] A very important contribution to models of recombination comes from Radding[31] (for review, see References 32 and 33) (Figure 1) and is introduced in detail in Chapter 2. Nicks on two homologous double-stranded molecules initiate the strand exchange of homologous duplex DNAs. The subsequent resolution of intermediate Holliday structures leads to newly arranged heteroduplexes. This model and its variations base on the discovery of recA in *E. coli.*[34,35] This protein promotes the pairing of homologous DNA molecules.[36] Meanwhile, equivalent proteins and activities have also been discovered and described in lower and higher eukaryotes.[37-40] According to the Meselson/Radding model, which is also called the Aviemore model,[31] recA promotes a pairing of an at least partially single-stranded DNA with duplex DNA in three steps. In a presynaptic phase, the protein binds to single-stranded DNA and polymerizes to a nucleoprotein filament. In the following synaptic phase, this filament binds non-specifically to duplex DNA and migrates along the DNA until a homologous region is found. *In vitro* this homologous pairing is initiated by DNA synthesis.[41] In the last step, the actual strand exchange occurs.

Another, variant model to explain homologous recombination is the double strand break model, initially mentioned by Resnick[42] and later modified by others[43-46] (Figure 4). A model that leaves one recombinant duplex, and as an intermediate, two DNA ends (which have to be closed in a second round of half cross-over), is called sequential half reciprocal recombination[5] (Figure 5).

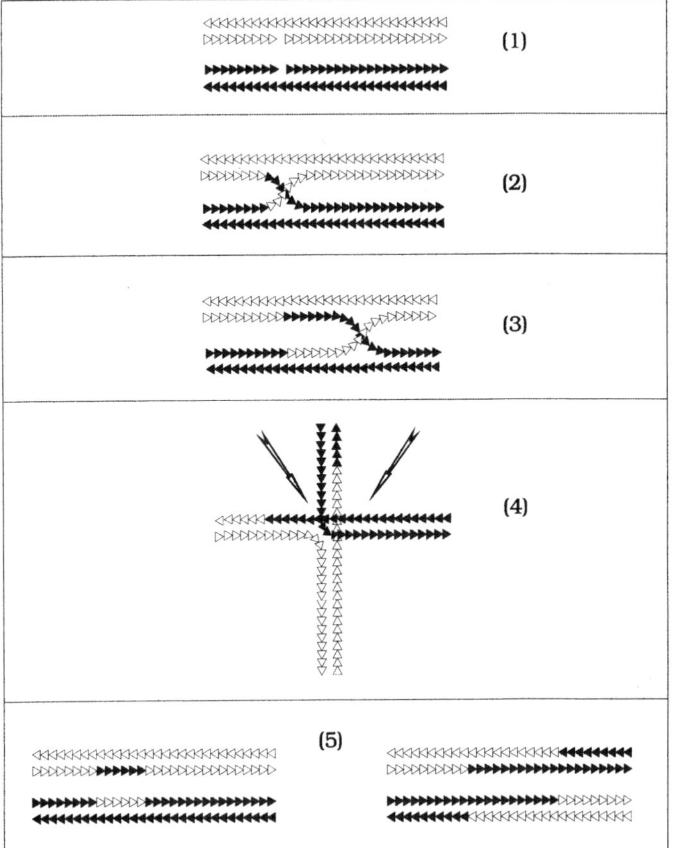

***Figure 3**. The Holliday model. An internal nick **(1)** in the two participating double-stranded molecules (white and black triangles) leads to a local strand exchange **(2)**. This point of exchange can migrate (branch migration) **(3)**. A different view of this migration product **(4)** shows how this Holliday structure can be resolved. The arrows point to one set of potential cutting and rejoining sites. The two reciprocal products that result are given in [**(5)** left, right]. Using alternate cleaving and rejoining sites, by twisting the molecule **(4)**, yields a different set of products. Further processes will repair differences in complementary strands.*

The double strand break model in fungi[43,47] and prokaryotes[2] has been reviewed extensively and is in its early development based on work of Stahl[48,49] in prokaryotes, in yeast,[15,50] and in higher eukaryotes.[42] An exonuclease cleaves DNA; another enzyme has DNA annealing activity. The initial breakpoint is expanded to a gap. These ends invade a homologous part of the DNA and are used either as primers for synthesis or trigger an exchange, leaving the other ends reactive and causing a cross-over.

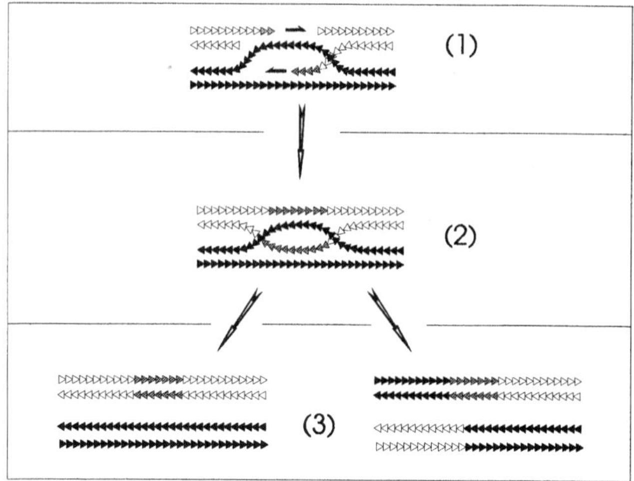

Figure 4. The double strand break model for homologous recombination. According to this model the exchange of genetic information begins with a double strand break of one partner (white triangles in (1)), which is expanded to a gap by exonucleases. The following strand displacement (black triangles in (1)), enables the reconstitution of that part of the sequence lost in the gap by DNA polymerase I extension. After ligation to their original strands a double Holliday structure **(2)** results. The resolution of this Holliday structure can [**(3)** right] or cannot [**(3)** left] involve cross-over.

Currently, there are two strand break models. In one model, a single-strand cut is introduced in each DNA duplex. The strands are exchanged to form a Holliday structure, and this structure is then resolved by cleavage and exchange of the other pair of strands.[51,52] Depending on the direction and extent of strand rotation preceding the resolution of the Holliday structure, these exchanges of single strands can lead to different products. This model bears resemblance to the half crossing over model,[5] which also leaves one or two ends at the end of the first round after two parental duplexes generate one recombinant duplex and reactive ends, which initiate a second round of half crossing over. If one of these ends engages in a second site cut during the cross-over at a short distance from the first exchange, the product resembles a gene conversion, and the half cross-over results then in a half reciprocal recombination. There are actually situations when one round is sufficient, such as in the case of telomere conversion in Trypanosoma.[5,53]

In the second model, double strand cuts are made at both parental DNAs, and double strands are exchanged by single rotation before resolution of the intermediate Holliday structure and ligation.[51]

B. Novel mechanisms

Recently a three-stranded intermediate structure[54] (Figure 6) was proposed by several groups[55] to facilitate homologous recombination. In this structure,

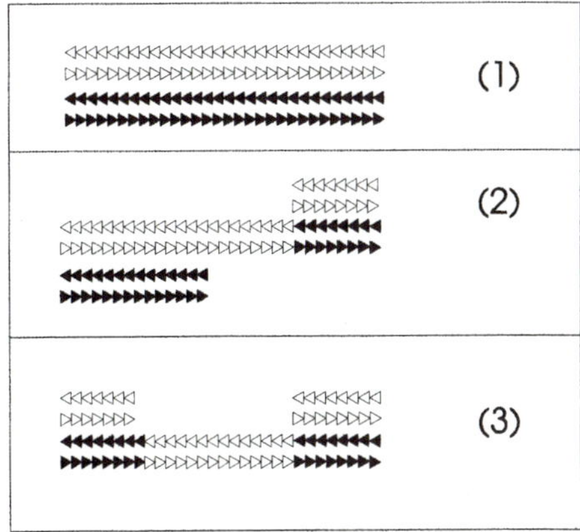

Figure 5*. The principle of the half crossing-over model. Two intact double-stranded DNA molecules* ***(1)*** *lead to one intact recombinant molecule and to two ends* ***(2)****. The ends engage in a second half cross-over at a different site, probably due to exonucleolytic digestion of free DNA ends in* ***(3)****. The resulting double-strand gap in the "white" DNA molecule can be repaired in different ways (see also Figure 4).*

the invading single-stranded DNA is interacting with the major groove of the duplex DNA. Homology-specific strands in the triple helix are in parallel orientation. Four-stranded DNA could also be involved in the pairing of recombination.[55]

There are also other mechanistic possibilities leading to homologous recombination and/or gene conversion that should be considered. A recombination of RNA with RNA during viral replication, or by using the splicing apparatus or the recombination of RNA with DNA during reverse transcription of, e.g., endogenous retroviruses and retroposons, could lead to homologous recombination or gene conversion. RNA could also be used as template to close DNA double strand breaks in analogy to DNA as depicted in Figure 6.

Generally, it is assumed that recombination, particularly homologous recombination, is taking place among DNA molecules. However, a number of observations indicate that RNA is also a suitable template for recombination. Obviously, a recombination of RNA molecules occurs during splicing, although this does not require homologies of the newly joint products either for the normal *cis*-splicing or for the seldom observed *trans*-splicing.[56,57] Mechanisms of homologous recombination among RNAs have recently been described during reverse transcription of retroviral RNA.[58] It is assumed that the reverse transcriptase switches templates during the synthesis of the minus strand, when the newly generated minus-strand DNA has a sufficient homology to the plus-strand template. The argument is that the recombination occurs only after packaging of these different RNAs into viral particles. However, a direct recombination between double-stranded RNAs by a mechanism similar to those models introduced for DNAs (see above) would need complementary

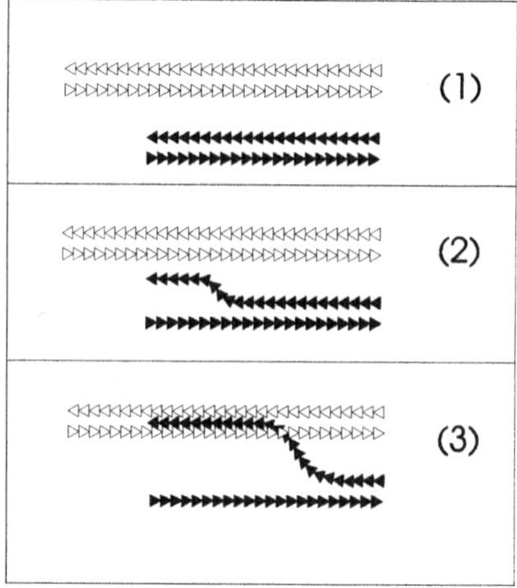

Figure 6*. Some steps in triple strand formation. A "black" linear DNA fragment interacts with a white intact target* ***(1)****. In a synaptic reaction one strand of the homologous fragment anneals to the double-stranded molecule* ***(2)****. The resulting triple strand* ***(3)*** *is resolved either by reversion of the process or by interaction of one of the "white" target strands with its complementary single strand of the "black" DNA.*

strands to interact because of their homology. That would imply that a direct interaction of complementary RNA in the absence of reverse transcriptase is also possible. This view of recombination among RNAs is supported by recombination observed in coronavirus,[59] which do not have a DNA intermediate, but have complementary RNAs in the cytoplasm. Recombination among at least partially homologous RNAs has also been described in other RNA viruses.[60,61] At least in those cases presented here, there are indications that a copy-choice mechanism, i.e., an alternate use of templates, involves RNA polymerase.[62] Another model, called RNA-mediated recombination,[6] indicates that RNA serves as an intermediate in recombination through a reverse transcriptase in lower and higher eukaryotes independent of viral infections. This model is used to explain recombination between RNAs, specifically reverse transcripts and homologous sites in the genome, the precise removal of introns,[64] and gene conversion of dispersed repeated sequences.[65] In yeast, this recombination requires expression of the retroposon Ty. In mammals, the expression of an analogous retroposon, LINE (Ll),[66] might supply the necessary enzymatic activity. To explain the precise loss of an intron,[64] one could also assume that a process similar to the above introduced three-strandedness leads to the use of RNA as a template to close a double strand gap (see Figure 6).

These observations indicate that even mutational processes that lead, for instance, to hypervariability of certain sequences or hot spots of mutation involve RNA during recombination. The idea is that RNA is the primary molecule that, due to faulty transcription, acquires mutations, which then (in a mechanism that is similar to gene conversion) introduces new sequence variations into the genome. Due to the enormous amount of RNA molecules generated per cell, mutational events in RNA are much more frequent than in newly synthesized DNA molecules. The reimport of such mutations could be a plausible way to ease adaptation processes.

This sheds a completely new light on the role of endogenous retroposons and retroviruses, especially considering the result that the presence of a part of the repeated element DRE increases the rate of homologous recombinations in *Dictyostelium.*[67,258] Although probably for a different reason, hypervariable minisatellite DNA also favors targeted events and is described as a hotspot for homologous recombination (in human cells).[68] Not only do retroposons seem to have a high homology even among non-functional copies,[66,69] but, due to their reverse transcriptase activity, they also seem to play a role in repair and evolutionary adaptation. Therefore, such elements might be of use to increase the inherently low efficiency of mammalian gene targeting attempts.

IV. Gene targeting

These mechanisms have been studied mainly in prokaryotes and lower eukaryotes[70] (see Chapter 2 and Chapter 3), but these models are also applicable on higher eukaryotes.

A. Gene disruption and the yeast system

Of all eukaryotic systems, recombination events are probably studied best in yeast. Its genome is readily accessible to genetic manipulation using plasmid technologies,[70,71] and even the technique of gene disruption has been established and reviewed quite extensively.[70] The enormous potential of this latter method includes the alteration or deletion of specific regions and the determination whether cloned fragments contain a specific gene and if this is essential. Using a straightforward, one-step technology, gene disruptions can be achieved with simple standard plasmid manipulations and are quite reproducible. A large number of yeast genes has been identified and characterized using cloning techniques, either by selecting clones from gene banks by hybridization with RNA or RNA-derived probes[72-75] or by using homologies of conserved regions to analogous regions of defined genes of other species.[76,77] Another way to select clones uses complementation in yeast.[78-80] Yeast mutants that lack certain functions reacquire those functions upon transformation with the corresponding DNA fragment. Complementation of mutations does not necessarily identify a wild type gene, but may also indicate a phenotypic suppression of that gene. The integration of the cloned gene at the genetic locus with a close linkage to the investigated gene is not sufficient proof that it is not a tightly linked suppressor. A gene disruption, however, unambiguously proves the identity of a cloned fragment. However, the integration of a disrupted gene into the corresponding chromosomal region creates a functional mutant in a wild-type strain. This allows the study of its function and shows genetic linkages by tetrade analysis. Other gene disruption experiments are based on a method by Scherer and Davis,[81,82] which they called transplacement, or on a method by Shortle et al.[83] called integration disruption. Transplacement introduces an altered cloned sequence in a plasmid by homologous recombination into the chromosome. The newly integrated plasmid sequences are now flanked by an altered and a wild-type copy of the gene. In a second step, the duplicated regions recombine and the plasmid and wild-type sequences are excised. Obviously, the major disadvantage of that method is that a series of recombinational events must take place and a large number of recombinants must be examined in order to detect the correct replacement. Integration disruptions make a mutant phenotype by insertion of a non-fractional fragment of the investigated gene, resulting in two mutated copies of that gene, one copy deleted at the 5′ end and the other one at the 3′ end. Only fragments of sufficient length can be used, but should not be too long since that could lead to a recombinational reversion of the integration step.

These and other molecular, biological tools have lead to an overwhelming number of publications on yeast recombination in recent years, reviewed in Petes et al.[84] Novel assays and novel mutants have been developed, and the function and the biochemistry of a huge number of gene products have been studied. This led to the establishment of new and better models of recombination in yeast backed by the characterization of recombinational intermediates. A number of new mutants has been characterized and investigated. For instance, *S. cerevisiae* spol3 mutants bypass meiosis I and, thus, enable the classification of mutants as early or late.[85] Recently, therefore, rad 52 was determined as a mutation with a late function in meiosis, and REC102 has an effect on the synaptonemal complex (SC).[86] This latter mutation is suppressed by the spol3 mutation, and the double mutation shows no meiotic recombination. Another gene with a function in early meiosis is ME1487, which is under mating type and additionally under nutritional control (conditions to induce meiosis). The lack of meiotic recombinations is due to defective SC formation: without chromosome pairing, recombination is not initiated. Recently, a gene that is responsible for the resolution of cross-overs by cleaving Holliday junctions, a cruciform-cutting endonuclease (CCEl), has been identified.[88] It is located on chromosome 11. Since there is still a residual cruciform-cutting activity, the existence of at least one additional protein with a similar cleavage function is assumed. Also, the gene responsible for a strand exchange (SEP 1, identical to DST 2) has been cloned, sequenced, and further analyzed.[89,90] The protein is active in strand exchange and has a 5′ to 3′ nuclease activity, thus conferring a 3′ to 5′ directional strand exchange reaction.[91] The protein shows no homologies to other known proteins, and its disfunction results in a low meiotic intragenic recombination, as does another strand transfer protein, DST 1.[92] Single-stranded DNA (ssDNA) binding activity could be attributed to RAD10, which catalyzes the reassociation of denatured double-stranded DNA (dsDNA).[93] Studies on low and high post-meiotic segregation alleles in tandem array and their recombination rates suggest an average excision tract length of 500 to 900 bp.[94] Studying the meiotic recombination initiator site at ARG4, 3′ extended ssDNA flanks of the initiation site showed a length of up to 800 bases, indicating a 5′ to 3′ exonuclease digestion from the site of a double strand break (DSB). Since a rad 50S mutant showed more DSB, but no ssDNA, the authors assume gene conversion as a result of mismatch repair.[95]

In order to define their function in a well-described and confined system, and to eventually use them as novel tools in recombination, foreign recombinases have been expressed in yeast. Especially noteworthy is the cre-loxP system of Sauer.[96] The principle of this technique is to bring two sequence-specific recombinase recognition sites, loxP sites, together and to promote a cross-over at these sites. The cre-encoded product, a site specific recombinase, does this with excellent efficiency. Under experimental conditions with only one loxP site,[97] the resulting inversion and translocation rearrangements are still specific.

The product of gene II of the bacterial phage fl sets normally site- and strand-specific nicks, and in yeast containing a gene II recognition site, actually leads to a raised level of gene conversions and cross-overs.[98]

A nuclease encoded by the yeast HO-locus induces site-specific double strand breaks. Placing this gene under galactose-dependent regulation, the genetics of the repair of these double strand breaks could be studied in a plasmid carrying a recognition site between direct repeat. Two independent pathways are suggested: (1) duplication is retained and (2) recombination results in a single copy. Sugawara and Haber[99] located the initiation of the cut by the HO-encoded nuclease between direct repeats. They observed 3′ ssDNA on both sides of the cleavage site suggesting a 5′ to 3′ exonuclease activity acting bidirectionally. Interestingly, they found a length dependence for intramolecular recombination. More than 400 bp of duplication is used efficiently, and the minimum sequence used successfully (with about 3% recombinations) is 89 bp. Especially in yeast, the role of retroposons and their reverse transcriptase has been studied extensively. The first indication of RNA involvement in yeast recombination was the loss of mitochondrial introns.[100] This observation was used to explain the origin of processed pseudogenes and the participation of RNAs or cDNAs in gene conversion events, as indicated by experiments of Derr et al.[63] A role for Ty-elements in these events was indicated by Ty sequences flanking the pseudogenes after recombination and a dependence on the SPT 3 gene. Others could show that Ty elements undergo (through their transcription and reverse transcription) gene conversion with other genomic Ty-elements.[101]

B. Gene targeting in other systems

Yeast, although a very important and powerful system, is, of course, only one example of a large number of other organisms available for a targeted alteration of a specific gene in its genomic locus. Although it is possible to target most genes in yeast and, thus, to analyze them on a genomic level, there are some drawbacks: yeast is not a useful system for the study of developmental phenomena or cell-cell communication.[102] Other systems also use lower eukaryotes[103] such as filamentous fungi, including *Aspergillus nidulans,*[104,105] *Neurospora crassa,*[106] *Podospora anserina,*[107] *Coprinus cinereus,*[108] and *Ustilago maydis,*[109] which appears to be of special importance for its enzyme systems involved in recombination, recl and rec2[38,110] which are relatively well studied. *Dictyostelium discoideum* is also a very important system where homologous recombination occurs with high frequency.[111] The main caveat of *D. discoideum* is that it has no natural sexual cycle, which limits experiments to the use of parasexual genetic techniques and hampers the analysis of genetic manipulations. Furthermore, several ciliated protozoa, including *Paramecium tetraurelia,*[112] *Tetrahymena thermophila,*[113] and *Stylonchia mytilus,*[114] are also included among the species suitable for homologous recombination studies. Targeting experiments with *Trypanosoma sp.* are discussed in Chapter 3.

Of the recombination systems in higher metazoa, I would like to direct attention to *Drosophila melanogaster*, which easily transforms with vectors based on the transposable P-element[115,116] and is also accessible for homologous recombination.[117] Another interesting and genetically very well-defined system is the nematode *Caenorhabditis elegans,*[118] which may also be used for homologous recombination experiments.[119] Interestingly, recent progress has been made in using transposon-based methods to target specific genes in the genomes of both animal systems (*D. melanogaster* and *C. elegans*). In *D. melanogaster,* the transposable P-elements have been modified and developed as transformation vectors and enhancer traps.[120-122] These modified P-elements express high levels of transposase and are unable to transpose,[123] but can activate P-elements with internal deletions or replacements. The system makes use of the fact that P-element transpositions induce gap repair (similar to Figure 4) and that this gap can be repaired by homologous recombination.[124] Similar experiments on germline excision that used the transposable element Tcl were performed in *C. elegans.*[119, 125]

Of plant systems, which are discussed in detail in Chapter 4, I would only like to point out the unicellular green alga *Chlamydomonas reinhardii*, since it was used for transformation with DNA-loaded microprojectiles very early,[126] a transfection method that has been established for other plant systems as well. Vertebrate systems will be discussed in detail in the following parts of this chapter and in Chapters 5 and 6.

C. Early studies in mammalian cells with artificial targets

The relatively high rate of homologous recombination in yeast has not only helped to define a number of parameters necessary for successful targeting and to develop recombination models for lower eukaryotes, but these findings also encouraged attempts on other, higher eukaryotic systems. Gene targeting in some of these systems is discussed in detail in later chapters. We will, therefore, focus here primarily on early studies in mammalian cells and later on more sophisticated studies.

As expected, genes in genomes of mammalian cells are more difficult to target than in yeast. Not only is their genome size (with more than 10^9 bp) considerably larger than that of yeast, these cells also seem to be quite efficient in random integration of transfected DNA.[127] For these and other reasons (e.g., the difficulty of handling and transfecting mammalian cells), a rather low success rate for gene targeting was expected. To facilitate the recovery of rare targeting events, early studies used artificially introduced loci with a defective marker gene to make them selectable targets. Also, these loci would be available for targeting in larger numbers than natural loci. Defective markers that were used were genes such as the neomycin-resistance gene[128-130] or the herpes TK gene.[131] These marker constructs were stably integrated into a chromosome. The resulting cell line was then transfected with a second copy of the marker gene harboring a different mutation. The selectable reconstruction of a functional marker indicated a successful targeting event. In these initial studies, the ratio of gene targeting to random integration was between 10^2 to 10.[5] This was very encouraging, and at the same time, proved that gene targeting in mammalian cells was a procedure that might be quite efficient as soon as more parameters of the underlying process were evaluated. Variations in the targeting efficiency from one cell line to another could influence the rate of gene targeting,[130,131] as well as the local chromatide structure of the integration site of the first, defective, artificial, selectable marker gene. Another very important parameter seemed to be the method applied to introduce exogenous DNA. Thomas et al.,[132] for instance, reported a far better ratio of targeted vs. random integration than Lin et al.[131] (see also below). When looking for more potentially rate-limiting steps, neither the number of targets[133] nor the amount of the incoming DNA[132,134] appeared to change the homologous integration efficiency (see below: Dependence on copy number). Therefore, the initial interaction between incoming and chromosomal DNA was not considered rate limiting, implying that random integration does not compete with targeting, i.e., two different independent systems would be involved (see below: Improving efficiency). Surprisingly, when targeting a locus with 800 copies, only a single targeting event typically occurs,[133] suggesting that one targeting event may exclude additional targetings.

D. Studies in mammalian cells with endogenous targets

There are two main differences between the recombinational behavior of prokaryotes and lower eukaroytes, mainly *E. coli* and yeast, and higher eukaryotes, namely mammalian systems: the extent and the frequency of recombinations in those different systems. Exogenous, incoming DNA recombines frequently and quickly upon entry into mammalian cells. However, DNA sequences stably integrated into the chromosome, unlike lower organisms, have a very low recombination rate.[135]

Another difference is that in higher eukaryotes the ratio of homologous vs. non-homologous recombination is getting increasingly smaller, at least under *in vitro* conditions.[127] An exception is the recombination between two exogenous DNA fragments. Here, very efficient ligation of incoming linear DNA to concatamers has been demonstrated.[37,64,136] Therefore, we have to discriminate among interactions of two exogenous, incoming DNAs, interactions of two sequences stably integrated in the chromosomal background, and interactions of exogenous and endogenous, chromosomal DNA.

For our purpose, the analysis and discussion of gene targeting, we will focus on the latter case (exogenous and chromosomal DNA interaction), particularly in higher eukaryotic systems.

The term homologous recombination is commonly interchanged with the term gene targeting in current literature, although it has a broader meaning. The enormous number of those publications in recent years reflects the importance that the development of such a technology will have. Whether it is used for insertion or replacement, knockout (disruption-of-function) or correction of defects, the main problems with gene targeting in complex systems are still its inefficiency, its fidelity, and the frequency of homologous vs. random integrations.

V. Parameters of efficiency

Most integrations of transfected DNA molecules into genomic DNA will normally be random (illegitimate recombination) with a success rate of anywhere between 10^{-1} to 10^{-5}.[132,137,138] This is quite different from the fate of transfected DNA in lower eukaryotes, where targeting is the normal event and illegitimate recombination the exception.[127] It is also different from the homologous recombination rate of two incoming DNA fragments.[139] Likewise, interchromosomal homologous recombination also seems to be of great fidelity in mammalian cells.[140] This indicates, however, that the size of the mammalian genome, with several 10^9 bp, is not the only parameter to be considered, although it was probably one reason why the early attempts at gene targeting in higher systems not only made use of selectable targets, but also involved artificial targeting loci rather than naturally occurring loci. Even earlier attempts used two incoming DNAs as targets for each other. After introduction of a selectable construct with a defective marker, its stable integration was tested. Such a cell line was then transfected with a second copy of that marker, but with a different mutation. The most common markers were G418-resistance (neomycine-resistance gene) and hypoxanthin-aminopterin selection (Herpes TK gene), with recombination rates of 10^{-2} to 10^{-5}.[128,130-132] The importance of double strand breaks was recognized very early,[141] but it was also suggested that local chromatin structures and local variations in accessibility and functional levels of DNA region were responsible for those fluctuating rates.[131,132] In addition, the method by which the DNA was introduced seemed to have a major effect on homologous integration rates (see below: Ways of introducing DNA). For instance, Thomas et al.[132] observed a ratio of homologous vs. random integration of 1:100, with a total of 1 targeted recombinant per 1000 cells when using microinjection. Obviously, some ways of transfecting DNA damaged the exogenous DNA more severely than others.[142,143] Another theory to explain why the transfection and recombination rates differed is that the incoming DNA lacks protection by nucleic acid-binding proteins and that even the accessibility of the DNA by such proteins varies for the different methods. That could explain why methods such as $CaP0_4$- or DEAE-mediated transfection often showed lower rates than ways releasing "naked" DNA into the nuclei, such as microinjection or electroporation.[144,145] This could also explain why methods employing viruses[146,147] or viral capsids[148,149] generally tend to show a reasonable recombination efficiency. This may also be the reason why the "flow" of the genetic information is generally from the exogenous to the genomic DNA, and a flow from the chromosomal to the transfected DNA is not as frequently observed,[150-152] although, recently, mechanisms for such observations have been proposed.[153,154] Others suggested that the direction of flow of genetic information would be predominantly from a longer to a shorter variant of the homologous sequences, so that it would be easier to produce an insertion than a deletion.[129] When natural loci, selectable[131,155,156] or non-selectable,[157–159] were used, similar efficiencies with rates of homologous recombinants of up to one in 4×10^{-7} were observed. A number of parameters would be expected to influence the targeting efficiency: the number of copies introduced or the number of endogenous targets; the length of the transfected fragments; the length of the homology region; and the degree of homology of the transfected DNA, including the length of interruptions of this homology region. Other parameters, such as local sequence peculiarities, e.g., (TG)n repeats,[160-163] might also promote reciprocal exchange.

VI. Dependence on copy number

Some tests of varying copy numbers, either of incoming DNA or of target loci, resulted in observations that might yield a better efficiency of gene targeting. Thomas et al.[132] injected increasing amounts of DNA into recipient nuclei (from 5 to 100 copies). This increase of the number of incoming DNA molecules did not improve the targeting efficiency.[132,134] At first glance, this seems to be a surprising result; but considering the size of human nuclear DNA, with 6×10^9 bp, and of a typical target, with about 3×10^3 bp, we might be dealing with differences in the ratio of incoming to chromosomal DNA that are too small to show a statistical significance. This implies that the initial interaction of exogenous with chromosomal DNA is not the rate-limiting step in a targeting reaction. Therefore, random integration does not seem to compete with homologous recombination mechanisms, although this has not been proven yet. Further reasons for this result might be that, as the amount of incoming DNA increases, the length of tandem arrays that are formed increases, probably due to the strong ligase activities in mammalian cells.[136,164] So the transfected DNA gets ligated and concatamerized faster than insertion into chromosomal DNA occurs. This is obviously detrimental to the homologous recombination rate, although the DNA ends are not always directly involved in the initial recombination event.[165] Additionally, it can be argued that as the copy number of potential competitor DNA, incoming DNA, increases, they not only compete for factors necessary for homologous recombination, but also recombine homologously among each other.[131,166,167] This is supported by the observation that the recombination among incoming DNA molecules is, in fact, dependent on DNA concentration.[167]

Another variable was tested in the work of Zengh and Wilson,[133] where different numbers of target sites of the DHFR gene were tested. This gene was present in the target cells in 1 to 400 copies in tandem array. However, these 400 copies still occupied only a relatively small region within the human chromosomal DNA, and unlike unlinked copies, did not give any additional entry sites for homology-driven pairing reactions. Also, targeting of tandem repeats may involve special mechanisms in order to remain stable in the chromosome, which again could influence its behavior in recombination reactions with exogenous DNA.

VII. Homology requirements

An early study of the influence of the degree of homology on the homologous recombination rate was performed by Thomas & Capecchi.[168] When they increased the length of homologous stretches from 4 to 9 kb, they found an overall increase of targeting efficiency by a factor of 10, but at the same time, the number of illegitimate recombinations rose by a factor of 40. It seems important to mention that Thomas & Capecchi used the HPRT gene in embryonal stem cells (ES-cells) as target, since results for varying conditions are not consistently comparable. Although others[169] reported similar rates for different genes and in different cells, Doetschmann et al.[137] achieved a higher success rate in a very similar system with an HPRT construct of less homology. Whether this is due to differences in the targeted area of the gene or experimental details in the mode of transfecting the exogenous DNA is not clear. Yet other studies showed a gradual dependence of homologous recombination rate with the shared homologies of the transfected DNA fragments.[170,171] Until recently, it was assumed, based partly on studies of prokaryotes,[172] but also other systems,[173] that the frequency of gene targeting was decreasing with increasing length of interruptions of the homologous sequence.[6,8,168] Mansour et al.,[174] however, showed that varying length (over 10000 bases) of interruptions showed the same rate on an otherwise identical experimental background. Even one-sided homology is reportedly sufficient for gene targeting.[175] Extrachromosomal recombination rates definitely were shown to depend on the length of the overlap.[173] One could conclude that as long as sufficiently long homologous regions are on both sides of such a non-homologous interruption, it is the number of such interruptions and their location within the homologous part, rather than their absolute length, which influence the targeting frequency. One would assume, supported by experimental approaches,[176] that a location close or at the end of a homology stretch initiates no or only one single cross-over event (leading to an insertion) rather than a double cross-over event, as necessary for replacement recombinations.[177] Although derived from an experiment on intrachromosomal recombination, Waldman and Liskay[176] determined a minimal uninterrupted length of homology at the termini of only 130 bp.

VIII. Fidelity of targeting reactions

The term fidelity of targeting reactions stands for an exact homologous recombination event with no secondary mutations to the genome. Since recombination involves breakage and repair, and eventually, gene conversion events (see above) with the excision of bases, one would expect gene targeting to be a mutagenic process. A further factor involved is the non-physiological presentation of the involved DNA. This might explain why intrachromosomal recombination shows a high fidelity,[140] but targeting events are often associated with mutations on the involved DNA,[129,137,178-181] although, recently, high-fidelity gene targeting has been described for ES-cells.[182] Small insertions,[129] deletions,[137] point mutations,[180] or aberrant recombinations[179,181] are the most frequently reported phenomena. The most likely reason for these faulty recombinations lies in the actual strand breaking and religation reaction, since the majority of these mutations is at or close to the actual integration site.

IX. Improving efficiency (and new approaches)

Applying all available knowledge about efficiency and fidelity of gene targeting, such as the length of homologous segments of the transfected DNA or its linearization, homologous recombinations are still relatively rare events. There seem to be several ways to favor targeted integration over random integration. One is derived from the observation that chromosomal DNA is different than transfected DNA, partly because it can be recognized as foreign due to a different methylation pattern or differences in the primary sequence. This has been observed in experiments with ES-cells to generate transgenic mice. When the targeted cells are from a different, non-isogenic mouse strain than the homologous transfected DNA, targeting efficiencies are lower.[138,183,184] Reports on the use of isogenic material in controlled experiments are rare and should be confirmed to rule out other parameters. Another difference of exogenous and endogenous DNA is that, unless packaged into viruses or viral capsids, the transfected

DNA has no, or at least not the correct, protein coating. Therefore, the incoming DNA is not accepted as a bonafide DNA by the recombination apparatus of the cell and will either be destroyed in the cell and/or the nucleus or be subject to some desperate repair attempts by alternative pathways. This may lead to random integration or concatamerization of fragments or may induce rearrangements and mutations. The direction to go, therefore, seems to be to use systems that coat DNA directly with protein, so it will not be recognized as foreign, or to artificially dock DNA to proteins that display the desired effects.

On the one hand random integration seems to be a last ditch solution for the repair apparatus of the cell;[17] on the other hand most recombinations between chromosomal sequences seem to be homologous (sister chromatide exchange, etc.). Therefore, it seems logical to block the "pathologic" repair mechanisms, while at the same time, leaving the homologous recombination mechanism intact. Since random integration of DNA is based mainly on end-joining,[127] which depends on intact ends of the DNA fragments while homologous recombination does not, Chang and Wilson[185] used DNA termini on their fragments, which they had modified to dideoxynucleotides. They could actually obtain evidence that this modification effectively blocked end-joining, but at the same time, still left some targeting activity. An alternate way to inhibit random integration is based on the observation that for this process, an enzyme, poly(ADP-ribosyl)transferase, is essential.[186,187] Consequently, some attempted to use this observation for gene targeting.[188] A disadvantage of this approach is that the drug used to inhibit poly(ADP-ribosyl)transferase, 3-methoxybenzamide (3-MB), is highly toxic and, therefore, of a very narrow therapeutic range. Attempts to replace this drug by more specific inhibitors, such as antisense-oligonucleotides or ribozymes that act on the mRNA of that enzyme, might yield a better efficiency for that concept.

These two pathways, random and homologous recombination, seem to be mutually exclusive, but the decision which of these pathways to use could also be predetermined by specific differentiation or cell cycle conditions. Surprisingly, however, while homologous integration has not been reported together with an additional random integration, single copy integrations at multiple sites have been observed when targeting a disrupted repetitive sequence.[189] These integrations result from multiple homologous recombination events in individual cells.[189] Further evidence that different enzymatic prerequisites are necessary for the random and homologous recombination pathway might come from the observation that, although nucleosomes on duplex DNA allow homologous pairing, they still prevent strand exchange (e.g., by recA).[190] While cross-over almost exclusively seems to happen in meiotic cells, which suggests that, at that time, factors are abundant that favor homologous recombination, it has been demonstrated[191] that homologous recombination between transfected DNA fragments occurs preferentially at early S-phase. So the synchronization of the cell cycle might also contribute to better efficiency of targeting. Meiosis is also a time when, apparently, other recombinationally active elements are activated whose contribution to homologous recombination are not yet understood. For instance, retroposons[257] and endogenous retroviruses[192,193] are activated in testicular environments where factors necessary for spermatogenesis and meiosis are available, similar to the transposable P-element of *Drosophila sp.*[124]

Another indication that the molecular and enzymatic requirements underlying homologous recombination are very effectively regulated and controlled is that one targeting event excludes additional targeting events (see Section IV. C). It is unclear why some loci[194-196] appeared to undergo targeting at higher frequencies than others. If this is due to sequence peculiarities, one could try to incorporate such sequences into constructs used for gene targeting, or if it is for the effect of specific recombinase systems, such as rag for T- and B-cells,[197] one could try to use these recombinase systems.

There are further differences between endogenous and exogenous DNA, such as the length of the involved DNA. It has been demonstrated that the length of the homology of the incoming DNA plays a crucial role,[137,168-170] but although the increase of efficiency seems to level off around 10 kb length,[168] it might still be possible to use extremely long DNA fragments, e.g., yeast artificial chromosomes (YACS).

X. Ways of introducing DNA

Even in early studies,[129,131,132] effects of different methods to transfect DNA on the overall recombination rate have been reported. While Thomas et al.,[132] who used microinjection,[198,199] observed a ratio of random targeting of about 10^{-2}, Lin et al.[131] reported a frequency of only about 10^{-5} using the so-called calciumphosphate coprecipitation method.[200-202] DNA forms long concatamers upon calciumphosphate coprecipitation[203] and may induce damage to the transfected DNA,[142,143] as do a number of other techniques.

A variety of techniques have been used to transfer new DNA sequences into higher eukaryotic cells *in vitro*. These include chemical methods[200,204,205] such as calciumphosphate coprecipitation or DEAE dextran mediated transfer,[204,206] and physical methods, e.g., electroporation[144,145,207] and microinjection.[198,199] Methods using biological vehicles appear to be the most gentle ones. These biological vehicles may be liposomes,[208-210] viral (DNA) and retroviral (RNA) vectors,[146,147] empty viral capsids,[149,211,212] or reconstituted viral proteins.[213,214] DNA transfected by electroporation or chemical techniques is often rearranged[136,215] and unstable in its integration[216] and, therefore, often yields a low efficiency in transformation.[164] Biological vectors that comprise viral DNA (or RNA) sequences may cause undesired effects; however, unlike all other methods, they do not transfect "naked" DNA, i.e., DNA coated with endogenous or viral proteins and, therefore, not shaped correctly to be accepted as a good substrate for recombination. A reason for a reverse flow of genetic information,[129,150-152] when incoming defective DNA is corrected or replenished by parts of the target sequence before it integrates into a random genomic locus, could be that the endogenous, correctly proteinized DNA is used to correct the deproteinized incoming DNA. A similar argument is that although recombination efficiency between chromosomal and incoming DNA seems to be independent of the number of incoming DNA fragments,[132,134] the rate of homologous recombination between transfected DNA molecules is dependent on DNA concentration.[167] We are trying to package linearized DNA fragments without viral sequences *in vivo* into viral capsules. This system has the safety of using DNA without viral sequences and, at the same time, should make use of the high transfection rate of viruses, e.g., the recently published Semliki system,[217] with a rate of one transfection per infectious unit. Of all the transfection methods described, only the particle-based systems (viral capsules, liposomes, reconstituted viral proteins, and viral vectors) are currently applicable for administration of DNA to living animals, although there are some more exotic methods[218-221] that also might be used to introduce DNA into adult animals and thus, ultimately, to perform somatic gene therapy. The efficiency of all the systems above is measured through many steps: the initial contact with the cell surface, the uptake through the membrane, the fate of the DNA in the cytoplasm (specifically in diverse compartments such as lysosomes on its way to the nucleus), the transport through the nuclear membrane, the reactions of the DNA with nuclear proteins, and finally, the actual recombination reaction. Many viruses have solved the initial problem and deliver their contents quite effectively and unaltered into the nucleus. The mimicking of such a viral system may induce a better overall transfection rate.

XI. Detecting recombinant cells in cell culture and in organisms

Due to the low frequency of gene targeting in higher eukaryotic cells, the screening for recombinants is a crucial step in detecting a cell with a desired modification. In a highly desirable situation, the targeting of a gene leads to a selectable phenotype in the host cell. However, only a very small fraction of genetic loci is selectable, and in many cases, the function of the target sequence is unknown. The best known, directly selectable gene is probably in the HPRT gene, which also has been used for gene targeting of embryonal stem cells or ES-cells.[156,168,222] Also, the manipulation of the HPRT gene results in a dominant change in male cells, since it is located on the X chromosome and the HPRT function can be used to select for HPRT-positive cells (by growing the cells in HAT-medium: Hypoxanthine, Aminopterin, Thymidine) and for HPRT negative cells (by growth in 6-thioguanidine).[156,168] Another method is to introduce positive selectable marker genes (included in the targeting DNA). Only three such selection markers have been used so far to do gene targeting in whole animals through ES-cells, the hygromycine B-resistance gene,[223] the neomycine-resistance gene,[168] and the HPRT-gene.[224] There are also methods that do not depend on target gene selection, but on an active transcription of the target site. In one method, a detectable product is generated and clones are selected by sib selection or replica plating.[225] If a defective gene that no longer expresses antigenic protein is corrected by homologous recombination with input DNA, only those cells that make the desired gene product will be identified by antibody screening. Another method uses two marker constructs: one that will integrate somewhere and is selectable and one that yields only positive cells when integrated homologously. This selection for the first selectable marker reduces the number of cells that will have to be screened by a second selection process.[157] Another method uses the transfer of a selectable marker gene. This marker, however, is only activated and transcribed if integrated in the target sequence.[178,181,226-228] A variation of this method not only uses the promoter and enhancer part of the target to yield a selectable product, but also results in a fusion protein that retains some functions of the target and, therefore, displays epitopes of both parts of the fusion protein. It is, therefore, called epitope addition.[229] There is another very efficient approach for reducing the number of non-homologous recombinants independent of gene function and gene expression. This procedure has been termed positive-negative selection (PNS).[230] This enrichment procedure consists of a positive selection for cells that have integrated into the construct of any site of the genome. These cells are then subjected to a second selection, discriminating all those that contain the construct at a random site. Another construct was designed[230] that contained parts of the homologous cellular gene, disrupted with a selectable (neomycin-resistance) gene, driven by a strong promoter. The linearized construct also contained, at its very end, a herpes simplex TK gene. If the input DNA inserted homologously as a gene replacement event, the TK gene would be lost and could be selected against with gancyclovir, since the endogenous, mammalian TK activity would not result in a toxic product. If the DNA, however, integrated and retained the TK part (e.g., because it had integrated randomly), i.e., if recombination occurred at and comprised the ends of the molecule,[127] their viral TK activity would kill the cell.

Another method that is independent of direct selection or expression of the target is based on Polymerase Chain Reaction (PCR). Kim and Smithies[231] first showed that novel sequence arrangements can easily be detected, even when the ratio of targeted to random is only about 1 in 10^6 cells[231] or if the amount of non-homologous DNA is as little as 20 bp[194,231] heterology. Obviously, PCR only amplifies predictable DNA fragments (which are only in the correctly recombined cells) by a factor of several hundred thousand to a million, but it does not enrich for a certain type of recombinants, and the few positive clones still have to be amplified.

XII. Applications to a large variety of genes and purposes

The technology of gene transfer opens a whole new era of applications. From a scientific standpoint, the analysis of gene functions and regulations seems to be of greatest advantage, since it provides the possibility of learning about and understanding genetic details and interactions that are otherwise inaccessible. Gene targeting will be an especially important tool in the genetic analysis of mammalian systems, e.g., by knockout experiments to assign functions to specific genes. The application of this methodology to the pluripotent ES-cells of, for example, mice, can also help to define the role of specific genes in the development and in the etiology of diseases. The easier targets will certainly be cells in culture, but a better understanding of the whole picture of living organisms can only be achieved through transgenic animals. By disabling a gene in knockout mice, for instance, we can learn about its structure and functions and can create animal models for genetic diseases. From other standpoints, the generation of modified novel food and drug products through genome manipulations appears attractive. In many cases, this is and will be done with constructs that insert randomly, but for many purposes, e.g., the eradication of a specific component in milk or special enzymes in fruits or crops, a targeted manipulation seems more feasible. In addition to these goals, a major justification for this technology comes from the medical applications. Not only will animal models help to provide cures for many diseases, genetic manipulation of the human body or parts thereof (also known as gene therapy) will be a favored treatment of the future. A targeted gene therapy where a majority of target cells is corrected at the site of the mutation, thereby reinstalling the original conditions, could actually lead to a "reinstitutio ad integrum" even for a patient suffering from a hereditary disorder, e.g., a specific defective enzyme.[13,232–234]

Application of the methodology of gene targeting to pluripotent mouse embryonic stem cells permits definition of the developmental or disease significance of any particular gene by generating the appropriate transgenic mouse. To introduce a planned alteration and a specific locus in the genome of a mouse, mouse embryonic stem cells (ES-cells) are first manipulated in a cell culture.[156,168,222,235] Cells containing the desired alteration are injected into recipient blastocystes. These blastocystes are reimplanted into pseudo-pregnant foster mothers. These techniques are very successful and are explained in detail by Hogan et al.[236] A successful manipulation yields chimeric pups, and if altered ES-cells contributed to the germline of such a chimera, its offspring will be a pure heterocygotes.[12] Although the first pluripotent embryonal carcinoma cell that could be used for reimplantation into blastocystes was developed in 1954,[237] the first ES-cells that were transmittable to the germline[235] and could be manipulated in culture[238,239] were isolated in 1981[240,241] from the inner cell mass of strain 138 and strain ICR mouse blastocystes. Culturing time and conditions[242,243] of these ES-cells seem to be crucial, as well as the strain of mice used as recipients.[244]

Although technically very difficult, a wide and continuously growing spectrum of genes has already been subjected to gene disruption. The c-myc protooncogene was truncated,[245] which leads to severe anemia due to a lack of erythropoetic and myeloid precursor cells. The gene for IgM heavy chain was a target,[246] which led to an absolute deficiency in B-cells and resulted in an immune deficiency. The knocked-out c-scr gene leads to osteopetrosis, the destroyed β2-microglobuline gene[247,248] leads to a deficiency in MHC class molecules and serves also to explain individual steps of T-cell maturation. The list of transgenic animals with disrupted genes is growing every day.

On the other major front of scientific progression, gene therapy, current approaches often use retroviral vectors to efficiently deliver functional genes.[249-251] Primarily, explantable and reimplantable tissue such as liver cells,[252] muscle cell precursors,[253] and bone marrow or bone marrow-derived cells[248] are targeted. But there are also reports where whole animals were subjected to a tissue targeting attempt.[148,254] Since the aspect of gene therapy is dealt with in detail in Chapter 8, I would like at this point to sidetrack to a different, transient approach of "gene targeting". The use of short DNA and RNA molecules with an antisense or ribozyme function has been reviewed before.[255,256] Correctly modified and galenically prepared, these genetic entities could help while they are needed and disappear later without leaving a permanent track. This concept could be used for fighting viral diseases or to inhibit activated oncogenes. In addition, because of their very high specificity, they would produce less side effects than chemotherapeutics that are used today for similar purposes.

This aspect of gene targeting, a nonpermanent modification of a specific gene in its genomic site, certainly does not lead to a permanent cure for a disease, but may, nonetheless, be preferred for its very transient nature, and shows yet another facet of the intriguing concept, the great capacity and potential of gene targeting.

References

1. Clark, A.J., Toward a metabolic interpretation of genetic recombination of *E. coli* and its phages, *Annu. Rev. Microbiol.*, 25, 437, 1971.

2. Smith, G.R., Homologous recombination in procaryotes, *Microbiol. Rev.*, 51, 4, 1988.

3. Hastings, P.J., Recombination in the eukaryotic nucleus, *Bioassays,* 9, 61, 1988.

4. Roeder, G.S., Stewart, S.E., Mitotic recombination in yeast, *Trends Genet.*, 4, 263, 1988.

5. Kobayashi, I., Mechanisms for gene conversion and homologous recombination: the double strand break repair model and the successive half crossing-over model, *Adv. Biophys.*, 28, 81, 1992.

6. Bollag, R.J., Waldman, A.S., Liskay, R.M., Homologous recombination in mammalian cells, *Annu. Rev. Genet.*, 23, 199, 1989.

7. Waldman, A.S., Targeted homologous recombination in mammalian cells, *Crit. Rev. Oncol. Hematol.*, 12, 49, 1992.

8. Frohman, M.A., Martin, G.R., Cut, paste, and save: new approaches to altering specific genes in mice, *Cell*, 56, 145, 1989.

9. Capecchi, M.R., The new mouse genetics: altering the genome by gene targeting, *Trends Genet.*, 5, 70, 1989.

10. Gridley, T., Insertional versus targeted mutagenesis in mice, *New Biol.*, 3, 1025, 1991.

11. Fung-Leung, W.-P., Mak, T.W., Embryonic stem cells and homologous recombination, *Curr. Opin. Immunol.*, 4, 189, 1992.

12. Koller, B.H., Smithies, O., Altering genes in animals by gene targeting, *Annu. Rev. Immunol.,* 10, 705, 1992.

13. Vega, M.A., Prospects for homologous recombination in human gene therapy, *Hum. Genet.*, 87, 245, 1991.

14. Hinnen, A., Hicks, J.B., Fink, G.R., Transformation of yeast, *Proc. Natl. Acad. Sci. U.S.A.,* 75, 1929, 1978.

15. Orr-Weaver, T.L., Szostak, J.W., Rothstein, R.J., Yeast transformation: a model system for the study of recombination, *Proc. Natl. Acad. Sci. U.S.A.*, 78, 6354, 1981.

16. Rothstein, R.J., One-step disruption in yeast, *Methods Enzymol.*, 101, 202, 1983.

17. Bohr, V.A., Evans, M.K., Fornace, A.J., Jr., DNA repair and its pathogenetic implications, *Lab. Invest.*, 61, 143, 1989.

18. Kucherlapati, R.S., Spencer, J., Moore, P.D., Homologous recombination catalyzed by human cell extracts, *Mol. Cell. Biol.*, 5, 714, 1985.

19. Hsieh, P., Meyn, M.S., Camerini-Otero, R.D., Partial purification and characterization of a recombinase from human cells, *Cell*, 44, 885, 1986.

20. Craig, N.L., The mechanism of conservative site-specific recombination, *Annu. Rev. Genet.*, 22, 77, 1988.

21. Porter, R.D., Modes of gene transfer in bacteria, in *Genetic Recombination*, Kucherlapati, R., Smith, G.R., Eds., American Society for Microbiology, Washington, DC, 1988, 1.

22. Echols, H., Gingery, R., Moore, L., Integrative recombination function of bacteriophage λ; evidence for a site specific recombination enzyme, *J. Mol. Biol.*, 34, 454, 1968.

23. Franklin, N.C., Illegitimate recombination, in *The Bacteriophage Lambda,* Hershey, A.D., Ed., Cold Spring Harbor Laboratory Press, Cold Spring Harbor, NY, 1971, 175.

24. Gottesmann, M.E., Yarmolinsky, M.B., Integration-negative mutants of bacteriophage lambda, *J. Mol. Biol.*, 31, 487, 1968.

25. Signer, E., Beckwith, J.R., Transposition of the lac region of *E. coli.* III. The mechanism of attachment of bacteriophage Φ80 to the bacterial chromosome, *J. Mol. Biol.*, 22, 33, 1966.

26. Griffith, R., The significance of pneumococcal types, *J. Hyg.*, 27, 113, 1928.

27. Whitehouse, H.L.K., A theory of crossing-over by means of hybrid deoxyribonucleic acid, *Nature,* 199, 1034, 1963.

28. Holliday, R., A mechanism for gene conversion in fungi, *Genet. Res.*, 5, 40, 1964.

29. Murchie, A.I., Clegg, R.M., von Kitzing, E., Duckett, D.R., Diekmann, S., Lilley, D.M., Fluorescence energy transfer shows that the four-way DNA junction is a right-handed cross of antiparallel molecules, *Nature*, 341, 763, 1989.

30. Sobell, H.M., Molecular mechanism for genetic recombination, *Proc. Natl. Acad. Sci. U.S.A.*, 68, 2483, 1972.

31. Meselson, M.S., Radding, C.M., A general model for genetic recombination, *Proc. Natl. Acad. Sci. U.S.A.*, 72, 358, 1975.

32. Radding, C.M., Homologous pairing and strand exchange in genetic recombination, *Annu. Rev. Genet.*, 16, 405, 1982.

33. Radding, C.M., Homologous pairing and strand exchange promoted by *Escherichia coli* recA protein, in *Genetic Recombination*, Kucherlapati, R., Smith, G.R., Eds., American Society for Microbiology, Washington, DC, 1988, 193.

34. Clark, A.J., Recombination deficient mutants of *E. coli* and other bacteria, *Annu. Rev. Genet.*, 7, 67, 1973.

35. Clark, A.J., Margulies, A.D., Isolation and characterization of recombination deficient mutants of *E. coli* K12, *Proc. Natl. Acad. Sci. U.S.A.*, 57, 451, 1965.

36. Cox, M.M., Lehmann, I.R., Enzymes for general recombination, *Annu. Rev. Biochem.*, 56, 229, 1987.

37. Fishel, R., Derbyshire, M.K., Moore, S.P., Young, C.S., Biochemical studies of homologous and nonhomologous recombination in human cells, *Biochimie*, 73, 257, 1991.

38. Kmiec, E.B., Holloman, W.K., Homologous pairing of DNA molecules by *Ustilago* Recl protein is promoted by sequences of Z-DNA, *Cell*, 44, 65, 1986.

39. McEntee, K., Weinstock, G.M., Lehman, I.R., RecA protein-catalyzed strand assimilation: stimulation by *Escherichia coli* single-stranded DNA-binding protein, *Proc. Natl. Acad. Sci. U.S.A.*, 77, 857, 1980.

40. Register III, J.C., Griffith, J., RecA protein filaments can juxtapose DNA ends: an activity that may reflect a function in DNA repair, *Proc. Natl. Acad. Sci. U.S.A.*, 83, 624, 1986.

41. Kodadek, T., Wong, M.L., Homologous pairing *in vitro* initiated by DNA synthesis, *Biochem. Biophys. Res. Commun.*, 169, 302, 1990.

42. Resnick, M.A., The repair of double-strand breaks in DNA; a model involving recombination, *J. Theor. Biol.*, 59, 97, 1976.

43. Szostak, J.W., Orr-Weaver, T.L., Rothstein, R.J., Stahl, F.W., The double-strand-break repair model for recombination, *Cell*, 33, 25, 1983.

44. Kobayashi, I., Murialdo, H., Crasemann, J.M., Stahl, M.M., Stahl, F.W., Orientation of cohesive end site cos determines the active orientation of chi sequence in stimulating recA, recBC-mediated recombination in phage lambda lytic infections, *Proc. Natl. Acad. Sci. U.S.A.*, 79, 5981, 1982.

45. Kobayashi, I., Stahl, M.M., Stahl, F.W., The mechanism of the chi-cos interaction in RecA-RecBC-mediated recombination in phage lambda, *Cold Spring Harbor Symp. Quant. Biol.*, 49, 497, 1984.

46. Rosenberg, S.M., Chi-stimulated patches are heteroduplex, with recombinant information on the phaqe lambda r chain, *Cell*, 48, 855, 1987.

47. Thaler, D.S., Stahl, F.W., DNA double-chain breaks in recombination of phage lambda and of yeast, *Annu. Rev. Genet.*, 22, 169, 1988.

48. Stahl, F.W., *Genetic Recombination*, W.H. Freeman and Co., San Francisco, 1979.

49. Stahl, F.W., Kobayashi, I., Stahl, M.M., In phage lambda, cos is a recombinator in the red pathway, *J. Mol. Biol.*, 181, 199, 1985.

50. Orr-Weaver, T.L., Szostak, J.W., Yeast recombination: the association between double-strand gap repair and crossing-over, *Proc. Natl. Acad. Sci. U.S.A.*, 80, 4417, 1983.

51. Hatfull, G.F., Grindley, N.D.F., Resolvases and DNA-invertases: a family of enzymes active in site-specific recombination, in *Genetic Recombination*, Kucherlapati, R., Smith, G.R., Eds., American Society for Microbiology, Washington, DC, 1988, 357.

52. Kikuchi, Y., Nash, H.A., Nicking-closing activity associated with bacteriophage int gene product, *Proc. Natl. Acad. Sci. U.S.A.*, 76, 3760, 1979.

53. Borst, P., Greaves, D.R., Programmed gene rearrangements altering gene expression, *Science*, 235, 658, 1987.

54. Rao, A.J., Dutreix, M., Radding, C.M., Stable three-stranded DNA made by recA protein, *Proc. Natl. Acad. Sci. U.S.A.*, 88, 2984, 1991.

55. Stasiak, A., Three-stranded DNA structure: is it the secret of DNA homologous recombination?, *Mol. Microbiol.*, 6, 3267, 1992.

56. Braun, R., Discontinuous RNA synthesis through *trans*-splicing, *Bioessays*, 5, 223, 1986.

57. Sutton, R.E., Boothroyd, J.C., Evidence for *trans* splicing in trypanosomes, *Cell*, 47, 527, 1986.

58. Stuhlmann, H., Berg, P., Homologous recombination of copackaged retrovirus RNAs during reverse transcription, *J. Virol.*, 66, 2378, 1992.

59. Liao, C.L., Lai, M.M., RNA recombination in coronavirus: recombination between viral genomic RNA and transfected RNA fragments, *J. Virol.*, 66, 6117, 1992.

60. Edwards, M.C., Petty, I.T., Jackson, A.O., RNA recombination in the genome of barley stripe mosaic virus, *Virology*, 189, 389, 1992.

61. Jarvis, T.C., Kirkegaard, K., Poliovirus RNA recombination: mechanistic studies in the absence of selection, *EMBO J.*, 11, 3135, 1992.

62. Kirkegaard, K., Baltimore, D., The mechanism of RNA recombination in poliovirus, *Cell*, 69, 433, 1986.

63. Derr, L.K., Strathern, J.N., Garfinkel, D.J., RNA-mediated recombination in *S. cerevisiae*, *Cell*, 67, 355, 1991.

64. Hunger-Bertling, K., Harrer, P., Bertling, W., Short DNA fragments induce site specific recombination in mammalian cells, *Mol. Cell. Biochem.*, 92, 107, 1990.

65. Derr, L.K., Strathern, J.N., A role for reverse transcripts in gene conversion, *Nature*, 75, 170, 1993.

66. Hutchison III, C.A., Hardies, S.C., Loeb, D.D., Shehee, W.R., Edgell, M.H., LINEs and related retroposons: long interspersed repeated sequences in the eucaryotic genome, in *Mobile DNA*, Berg, D.E., Howe, M., Eds., American Microbiology, Washington, DC, 1989, 593.

67. Schumann, G., Zundorf, I., Hofmann J., Marschalek, R., Dingermann, T., Expression of the retroposon DRE from *Dictyostelium discoideum*, submitted, 1993.

68. Wahls, W.P., Wallace, L.J., Moore, P.D., Hypervariable minisatellite DNA is a hotspot for homologous recombination in human cells, *Cell*, 60, 95, 1990.

69. Bertling, W., Full-length Ll retroposons contain tRNA-like sequences near the 5′-termini. Hypothesis on the replication mechanism of retroposons, *J. Theor. Biol.*, 138, 185, 1989.

70. Rothstein, R., Targeting, disruption, replacement, and allele rescue: integrative DNA transformation in yeast, *Methods Enzymol.*, 194, 281, 1991.

71. Orr-Weaver, T.L., Szostak, J.W., Rothstein, R.J., Genetic applications of yeast transformation with linear and gapped plasmids, *Methods Enzymol.*, 101, 228, 1983.

72. Kramer, R.A., Cameron, J.R., Davis, R.W., Isolation of bacteriophage lambda containing yeast ribosomal RNA genes: screening by *in situ* RNA hybridization to plaques, *Cell*, 8, 1976.

73. Woolford, J.L., Jr., Hereford, L.M., Rosbash, M., Isolation of cloned DNA sequences containing ribosomal protein genes from *Saccharomyces cerevisiae*, *Cell*, 18, 1247, 1979.

74. St. John, T.P., Davis, R.W., Isolation of galactose-inducible DNA sequences from *Saccharomyces cerevisiae* by differential plaque filter hybridization, *Cell*, 16, 443, 1979.

75. Holland, M.J., Holland, J.P., Jackson, K.A., Cloning of yeast genes coding for glycolytic enzymes, *Methods Enzymol.*, 68, 408, 1979.

76. Hereford, L., Fahrner, K., Woolford, J., Jr., Rosbash, M., Kaback, D.B., Isolation of yeast histone genes H2A and H2B, *Cell*, 18, 1261, 1979.

77. Gallwitz, D., Seidel, R., Molecular cloning of the actin gene from yeast *Saccharomyces cerevisiae*, *Nucl. Acids Res.*, 8, 1043, 1980.

78. Broach, J.R., Strathern, J.N., Hicks, J.B., Transformation in yeast: development of a hybrid cloning vector and isolation of the CANl gene, *Gene*, 8, 121, 1979.

79. Williamson, V.M., Bennetzen, J., Young, E.T., Nasmyth, K., Hall, B.D., Isolation of the structural gene for alcohol dehydrogenase by genetic complementation in yeast, *Nature*, 283, 214, 1980.

80. Nasmyth, K.A., Tatchell, K., The structure of transposable yeast mating type loci, *Cell*, 110, 753, 1980.

81. Scherer, S., Davis, R.W., Replacement of chromosome segments with altered DNA sequences constructed *in vitro*, *Proc. Natl. Acad. Sci. U.S.A.*, 76, 4951, 1979.

82. Winston, F., Chumley, F., Fink, G.R., Eviction and transplacement of mutant genes in yeast, *Methods Enzymol.*, 101, 211, 1983.

83. Shortle, D., Haber, J.E., Botstein, D., Lethal disruption of the the yeast actin gene by integrative DNA transformation, *Science*, 217, 371, 1982.

84. Petes, T.D., Malone, R.E., Symington, L.S., Recombination in yeast, in *The Molecular and Cellular Biology of the Yeast Saccharomyces*, Broach, J.R., Ed., Cold Spring Harbor Laboratory Press, Cold Spring Harbor, NY, 1991, 407.

85. Malone, R.E., Bullard, S., Hermiston, M., Rieger, R., Cool, M., Galbraith, A., Isolation of mutants defective in early steps of meiotic recombination in the yeast *Saccharomyces cerevisiae*, *Genetics*, 128, 79, 1991.

86. Bhargava, J., Engebrecht, J., Roeder, G.S., The recl02 mutant of yeast is defective in meiotic recombination and chromosome synapsis, *Genetics,* 130, 59, 1992.

87. Menees, T.M., Ross-MacDonald, P.B., Roeder, G.S., MEI4, a meiosis-specific yeast gene required for chromosome synapsis, *Mol. Cell. Biol.*, 12, 1340, 1992.

88. Zweifel, S.G., Fangman, W.L., A nuclear mutation reversing a biased transmission of yeast mitochondrial DNA, *Genetics*, 128, 241, 1991.

89. Tishkoff, D.X., Johnson, A.W., Kolodner, R.D., Molecular and genetic analysis of the gene encoding the *Saccharomyces cerevisiae* strand exchange protein Sepl , *Mol. Cel. Biol.*, 11, 2593, 1991.

90. Dykstra, C.C., Kitada, K., Clark, A.B., Hamatake, R.K., Sugino, A., Cloning and characterization of DST2, the gene for DNA strand transfer protein beta from *Saccharomyces cerevisiae*, *Mol. Cell. Biol.*, 11, 2583, 1991.

91. Johnson, A.W., Kolodner, R.D., Strand exchange protein 1 from *Saccharomyces cerevisiae*, a novel multifunctional protein that contains DNA strand exchange and exonuclease activities, *J. Biol. Chem.*, 266, 14046, 1991.

92. Clark, A.B., Dykstra, C.C., Sugino, A., Isolation, DNA sequence, and regulation of a *Saccharomyces cerevisiae* gene that encodes DNA strand transfer protein alpha, *Mol. Cell. Biol.*, 11, 2576, 1991.

93. Sung, P., Prakash, L., Prakash, S., Renaturation of DNA catalysed by yeast DNA repair and recombination protein RAD10, *Nature*, 355, 743, 1992.

94. Detloff, P., Petes, T.D., Measurements of excision repair tracts formed during meiotic recombination in *Saccharomyces cerevisiae*, *Mol. Cell. Biol.*, 12, 1805, 1992.

95. Sun, H., Treco, D., Szostak, J.W., Extensive 3′-overhanging, single-stranded DNA associated with the meiosis-specific double-strand breaks at the ARG4 recombination initiation site, *Cell*, 64, 1155, 1991.

96. Sauer, B., Functional expression of the cre-lox site-specific recombination system in the yeast *Saccharomyces cerevisiae, Mol. Cell. Biol.*, 7, 2087, 1987.

97. Sauer, B., Identification of cryptic lox sites in the yeast genome by selection for cre-mediated chromosome translocations that confer multiple drug resistance, *J. Mol. Biol.*, 91, 1, 1992.

98. Strathern, J.N., Weinstock, K.G., Higgins, D.R., McGill, C.B., A novel recombinator in yeast based on gene II protein from bacteriophage fl, *Genetics*, 127, 61, 1991.

99. Sugawara, N., Haber, J.E., Characterization of double-strand break-induced recombination: homology requirements and single-stranded DNA formation, *Mol. Cell. Biol.*, 12, 563, 1992.

100. Levra-Juillet, E., Boulet, A., Seraphin, B., Simon, M., Faye, G., Mitochondrial introns aIl and/or aI2 are needed for the *in vivo* deletion of intervening sequences, *Mol. Gen. Genet.*, 217, 168, 1989.

101. Melamed, C., Nevo, Y., Kupiec, M., Involvement of cDNA in homologous recombination between Ty elements in *Saccharomyces cerevisiae*, *Mol. Cell. Biol.*, 12, 1613, 1992.

102. Botstein, D., Fink, G.R., Yeast: an experimental organism for modern biology, *Science*, 240, 1439, 1988.

103. Hastings, P.J., Mechanism and control of recombination in fungi, *Mutat. Res.*, 284, 97, 1992.

104. Tilburn, J., Scazzocchio, C., Taylor, G.G., Zabicky-Zissman, J.H., Lockington, R.A., Davies, R.W., Transformation by integration in *Aspergillus nidulans*, *Gene*, 26, 205, 1983.

105. Johnstone, I.L., Transformation of *Aspergillus nidulans, Microbiol. Sci.*, 2, 307, 1985.

106. Case, M.E., Genetical and molecular analyses of qa-2 transformants in *Neurospora crassa*, *Genetics*, 113, 569, 1986.

107. Razanamparany, V., Begueret, J., Positive screening and transformation of ura5 mutants in the fungus *Podospora anserina*: characterization of the transformants, *Curr. Genet.*, 10, 811, 1986.

108. Binninger, D.M., Skrzynia, C., Pukkila, P.J., Casselton, L.A., DNA-mediated transformation of the basidiomycete *Coprinus cinereus*, *EMBO J.*, 6, 835, 1987.

109. Wang, J., Holden, D.W., Leong, S.A., Gene transfer system for the phytopathogenic fungus *Ustilago maydis*, *Proc. Natl. Acad. Sci. U.S.A.*, 85, 865, 1988.

110. Fotheringham, S., Holloman, W.K., Extrachromosomal recombination is deranged in the rec2 mutant of *Ustilago maydis*, *Genetics*, 129, 1052, 1991.

111. Spudich, J.A., Ed., *Dictyostelium discoideum*: molecular approaches to cell biology, *Methods Cell. Biol.*, 28, 1, 1987.

112. Godiska, R., Aufderheide, K.J., Gilley, D., Hendrie, P., Fitzwater, T., Preer, P.B., Preer, J.R., Jr., Transformation of *Paramecium* by microinjection of a cloned serotype gene, *Proc. Natl. Acad. Sci. U.S.A.*, 84, 7590, 1987.

113. Tondravi, M.M., Yao, M.C., Transformation of *Tetrahymena thermophila* by microinjection of ribosomal RNA genes, *Proc. Natl. Acad. Sci. U.S.A.*, 83, 4369, 1986.

114. Meyers, G., Helftenbein, E., Transfection of the hypotrichous ciliate *Stylonychia lemnae* with linear DNA vectors, *Gene*, 63, 31, 1988.

115. Rubin, G.M., Spradling, A.C., Genetic transformation of *Drosophila* with transposable element vectors, *Science*, 218, 348, 1982.

116. Steller, H., Pirrotta, V., Fate of DNA injected into early *Drosophila* embryos, *Dev. Biol.*, 109, 54, 1985.

117. Germeraad, S.E., Alteration of the gene for xanthine dehydrogenase in *Drosophila* following treatment with wild type DNA, *Mol. Gen. Genet.*, 199, 89, 1985.

118. Mello, C.C., Kramer, J.M., Stinchcomb, D., Ambros, V., Efficient gene transfer in *C. elegans:* extrachromosomal maintenance and integration of transforming sequences, *EMBO J.*, 10, 3959, 1991.

119. Rushforth, A.M., Saari, B., Anderson, P., Site-selected insertion of the transposon Tcl into *Caenorhabditis elegans* myosin light chain gene, *Mol. Cell. Biol.*, 13, 902, 1993.

120. Spradling, A.C., Rubin, G.M., Transposition of cloned P elements into *Drosophila* germ line chromosomes, *Science,* 218, 341, 1982.

121. Kane, C.J., Gehring, W.J., Detection *in situ* of genomic regulatory elements in *Drosophila*, *Proc. Natl. Acad. Sci. U.S.A.*, 84, 9123, 1987.

122. Bellen, H.J., Wilson, C., Gehring, W.J., Dissecting the complexity of the nervous system by enhancer detection, *Bioessays*, 12, 199, 1990.

123. Laski, F.A., Rio, D.C., Rubin, G.M., Tissue specificity of *Drosophila* P element transposition is regulated at the level of mRNA splicing, *Cell*, 44, 7, 1986.

124. Sentry, J.W., Kaiser, K., P element transposition and targeted manipulation of the *Drosophila* genome, *Trends Genet.*, 8, 329, 1992.

125. Moerman, D.G., Kiff, J.E., Waterston, R.H., Germline excision of the transposable element Tcl in *C. elegans*, *Nucl. Acids Res.*,19, 5669, 1991.

126. Boynton, J.E., Gillham, N.W., Harris, E.H., Hosler, J.P., Johnson, A.M., Jones, A.R., Randolph-Anderson, B.L., Robertson, D., Klein, T.M., Shark, J., Sandford, J.C., Chloroplast transformation in Chlamydomonas with high velocity micropro–jectiles, *Science*, 240, 1534, 1988.

127. Roth, D., Wilson, J., Illegitimate recombination in mammalian cells, in *Genetic Recombination*, Kucherlapati, R., Smith, G.R., Eds., American Society for Microbiology, Washington, DC, 1988, 621.

128. Smith, A.J., Berg, P., Homologous recombination between defective neo genes in mouse 3T6 cells, *Cold Spring Harb. Symp. Quant. Biol.*, 49, 171, 1984.

129. Thomas, K.R., Capecchi, M.R., Introduction of homologous DNA sequences into mammalian cells induces mutations in the cognate gene, *Nature*, 163, 34, 1986.

130. Song, K.Y., Schwartz, F., Maeda, N., Smithies, O., Kucherlapati, R., Accurate modification of a chromosomal plasmid by homologous recombination in human cells, *Proc. Natl. Acad. Sci. U.S.A.,* 84, 6820, 1987.

131. Lin, F.-L., Sperle, K., Sternberg, N., Recombination in mouse L cells between DNA introduced into cells and homologous chromosomal sequences, *Proc. Natl. Acad. Sci. U.S.A.*, 82, 1391, 1985.

132. Thomas, K.R., Folger, K.R., Capecchi, M.R., High frequency targeting of genes to specific sites in the mammalian genome, *Cell,* 44, 419, 1986.

133. Zheng, H., Wilson, J.H., Gene targeting in normal and amplified cell lines, *Nature*, 344, 170, 1990.

134. Rommerskirch, W., Graeber, I., Graessmann, M., Graessmann, A., Homologous recombination of SV40 DNA in COS7 cells occurs with high frequency in a gene dose independent fashion, *Nucl. Acids Res.*, 16, 941, 1988.

135. Subramani, S., Seaton, B.L., Homologous recombination in mitotically dividing mammalian cells, in *Genetic Recombination*, Kucherlapati, R., Smith, G.R., Eds., American Society for Microbiology, Washington, DC, 1988, 549.

136. Miller, C.K., Temin, H.M., High-efficiency ligation and recombination of DNA fragments by vertebrate cells, *Science*, 220, 606, 1983.

137. Doetschman, T., Maeda, N., Smithies, O., Targeted mutation of the Hprt gene in mouse embryonic stem cells, *Proc. Natl. Acad. Sci. U.S.A.*, 85, 8583, 1988.

138. Deng, C., Capecchi, M.R., Reexamination of gene targeting frequency as a function of the extent of homology between the targeting vector and the target locus, *Mol. Cell. Biol.*, 12, 3365, 1992.

139. Shimoda, K., Cai, X., Kuhara, T., Maejima, K., Reconstruction of a large DNA fragment from coinjected small fragments by homologous recombination in fertilized mouse eggs, *Nucl. Acids Res.*, 19, 6654, 1991.

140. Stachelek, J.L., Liskay, R.M., Accuracy of intrachromosomal gene conversion in mouse cells, *Nucl. Acids Res.*, 16, 4069, 1988.

141. Song, K.Y., Chekuri, L., Rauth, S., Ehrlich, S., Kucherlapati, R., Effect of double-strand breaks on homologous recombination in mammalian cells and extracts, *Mol. Cell. Biol.*, 5, 3331, 1985.

142. Lebkowski, J.S., DuBridge, R.B., Antell, E.A., Greisen, K.S., Calos, M.P., Transfected DNA is mutated in monkey, mouse, and human cells, *Mol. Cell. Biol.*, 4, 1951, 1984.

143. Wake, C.T., Gudewicz, T., Porter, T., White, A., Wilson, J.H., How damaged is the biologically active subpopulation of transfected DNA? *Mol. Cell. Biol.*, 195, 387, 1984.

144. Zimmermann, U., Electric field-mediated fusion and related electrical phenomena, *Biochim. Biophys. Acta*, 694, 227, 1982.

145. Neumann, E., Schaefer-Ridder, M., Wang, Y., Hofschneider, P.H., Gene transfer into mouse lyoma cells by electroporation in high electric fields, *EMBO J.*, 1, 841, 1982.

146. Hamer, D.H., Leder, P., Splicing and the formation of stable RNA, *Cell*, 18, 1299, 1979.

147. Gluzman, Y., *Eukaryotic Viral Vectors*, Gluzman, Y., Ed., Cold Spring Harbor Laboratory Press, Cold Spring Harbor, NY, 1982.

148. Bertling, W.M., Gareis, M., Paspaleeva, V., Zimmer, A., Kreuter, J., Nuernberg, E., Harrer, P., Use of liposomes, viral capsids, and nanoparticles as DNA carriers, *Biotechnol. Appl. Biochem.*, 13, 390, 1991.

149. Aposhian, H.V., Thayer, R.E., Qasba, P.K., Formation of nucleoprotein complexes between polyoma empty capsides and DNA, *J. Virol.*, 15, 645, 1975.

150. Jasin, M., de Villiers, J., Weber, F., Schaffner, W., High frequency of homologous recombination in mammalian cells between endogenous and introduced SV40 genomes, *Cell*, 43, 695, 1985.

151. Shaul, Y., Laub, O., Walker, M.D., Rutter, W.J., Homologous recombination between a defective virus and a chromosomal sequence in mammalian cells, *Proc. Natl. Acad. Sci. U.S.A.*, 82, 3781, 1985.

152. Subramani, S., Rescue of chromosomal T-antigen sequences onto extrachromosomally replicating, defective simian virus 40 DNA by homologous recombination, *Mol. Cell. Biol.*, 6, 1320, 1986.

153. Valancius, V., Smithies, O., Double-strand gap repair in a mammalian gene targeting reaction, *Mol. Cell. Biol.*, 11, 4389, 1991.

154. Aratani, Y., Okazaki, R., Koyama, H., End extension repair of introduced targeting vectors mediated by homologous recombination in mammalian cells, *Nucl. Acids Res.*, 18, 4795, 1992.

155. Adair, G.M., Nairn, R.S., Wilson, J.H., Seidman, M.M., Brotherman, K.A., MacKinnon, C., Scheerer, J.B., Targeted homologous recombination at the endogenous adenine phosphoribosyltransferase locus in Chinese hamster cells, *Proc. Natl. Acad. Sci. U.S.A.*, 86, 4574, 1989.

156. Doetschman, T., Gregg, R.G., Maeda, N., Hooper, M.L., Melton, D.W., Thompson, S., Smithies, O., Targeted correction of a mutant HPRT gene in mouse embryonic stem cells, *Nature*, 330, 576, 1987.

157. Smithies, O., Gregg, R.G., Boggs, S.S., Koralewski, M.A., Kucherlapati, R.S., Insertion of DNA sequences into the human chromosomal beta-globin locus by homologous recombination, *Nature*, 317, 230, 1985.

158. Baker, M.D., Pennell, N., Bosnaoyan, L., Shulman, M.J., Homologous recombination can restore normal immunoglobulin production in a mutant hybridoma cell line, *Proc. Natl. Acad. Sci. U.S.A.*, 85, 6432, 1988.

159. Baker, M.D., Shulman, M.J., Homologous recombination between transferred and chromosomal immunoglobulin k genes, *Mol. Cell. Biol.*, 8, 4041, 1988.

160. Treco, D., Arnheim, N., The evolutionarily conserved repetitive sequence d(TG,AC)n promotes reciprocal exchange and generates unusual recombinant tetrads during yeast meiosis, *Mol. Cell. Biol.*, 6, 3934, 1986.

161. Bullock, P., Miller, J., Botchan, M., Effects of poly [d(pGpT), d(pApC)] and poly[d(pCpG),d(pCpG)] repeats on homologous recombination in somatic cells, *Mol. Cell. Biol.*, 6, 3948, 1986.

162. Wahls, W.P., Moore, P.D., Homologous recombination enhancement conferred by the Z-DNA motif d(TG)30 is abrogated by simian virus T antigen binding to adjacent DNA sequences, *Mol. Cell. Biol.*, 10, 794, 1990.

163. Wahls, W.P., Wallace, L.J., Moore, P.D., The Z-DNA motif d(TG)30 promotes reception of information during gene conversion events while stimulating homologous recombination in human cells in culture, *Mol. Cell. Biol.*, 10, 785, 1990.

164. Folger, K.R., Wong, E.A., Wahl, G., Capecchi, M.R., Patterns of integration of DNA microinjected into cultured mammalian cells: evidence for homologous recombination between injected plasmid DNA molecules, *Mol. Cell. Biol.*, 2, 1372, 1982.

165. Hasty, P., Rivera-Pérez, J., Bradley, A., The role and fate of DNA ends for homologous recombination in embryonic stem cells, *Mol. Cell. Biol.*, 12, 2464, 1992.

166. Jasin, M., Liang F., Mouse embryonic stem cells exhibit high levels of extrachromosomal homologous recombination in a chloramphenicol acetyltransferase assay system, *Nucl. Acids Res.*, 19, 7171, 1991.

167. Waldman, A.S., Liskay, R.M., Differential effects of basepair mismatch on intrachromosomal versus extrachromosomal recombination in mouse cells, *Proc. Natl. Acad. Sci. U.S.A.*, 84, 5340, 1987.

168. Thomas, K.R., Capecchi, M.R., Site-directed mutagenesis by gene targeting in mouse embryo-derived stem cells, *Cell*, 51, 503, 1987.

169. Shulman, M.J., Nissen, L., Collins, C., Homologous recombination in hybridoma cells: dependence on time and fragment length, *Mol. Cell. Biol.*, 10, 4466, 1990.

170. Liskay, R.M., Letsou, A., Stachelek, J.L., Homology requirement for efficient gene conversion between duplicated chromosomal sequences in mammalian cells, *Genetics*, 115, 161, 1987.

171. Hasty, P., Rivera-Pérez, J., Bradley, A., The length of homology required for gene targeting in embryonic stem cells, *Mol. Cell. Biol.*, 11, 5586, 1991.

172. Watt, V.M., Ingles, C.J., Urdea, M.S., Rutter, W.J., Homology requirements for recombination in *Escherichia coli*, *Proc. Natl. Acad. Sci. U.S.A.*, 82, 4768, 1985.

173. Puchta, H., Hohn, B., A transient assay in plant cells reveals a positive correlation between extrachromosomal recombination rates and length of homologous overlap, *Nucl. Acids Res.*, 19, 2693, 1991.

174. Mansour, S.L., Thomas, K.R., Deng, C.X., Capecchi, M.R., Introduction of a lacZ reporter gene into the mouse int-2 locus by homologous recombination, *Proc. Natl. Acad. Sci. U.S.A.*, 87, 7688, 1990.

175. Berinstein, N., Pennell, N., Ottaway, C.A., Shulman, M.J., Gene replacement with one-sided homologous recombination, *Mol. Cell. Biol.*, 12, 360, 1992.

176. Waldman, A.S., Liskay, R.M., Dependence of intrachromosomal recombination in mammalian cells on uninterrupted homology, *Mol. Cell. Biol.*, 8, 5350, 1988.

177. Lee, M.-A., Kim, O.-J., Yates, J.L., Targeted gene disruption in Epstein-Barr, *Virology,* 189, 253, 1992.

178. Zijlstra, M., Li, E., Sajjadi, F., Subramani, S., Jaenisch, R., Germ-line transmission of a disrupted beta 2-microglobulin gene produced by homologous recombination in embryonic stem cells, *Nature*, 342, 435, 1989.

179. Sedivy, J.M., Sharp, P.A., Positive genetic selection for gene disruption in mammalian cells by homologous recombination, *Proc. Natl. Acad. Sci. U.S.A.*, 86, 227, 1989.

180. Brinster, R.L., Braun, R.E., Lo, D., Avarbock, M.R., Oram, F., Palmiter, R.D., Targeted correction of a major histocompatibility class II E alpha gene by DNA microinjected into mouse eggs, *Proc. Natl. Acad. Sci. U.S.A.,* 86, 7087, 1989.

181. Schwartzberg, P.L., Robertson, E.J., Goff, S.P., Targeted gene disruption of the endogenous c-abl locus by homologous recombination with DNA encoding a selectable fusion protein, *Proc. Natl. Acad. Sci. U.S.A.,* 87, 3210, 1990.

182. Thomas, K.R., Deng, C., Capecchi, M.R., High-fidelity gene targeting in embryonic stem cells by using sequence replacement vectors, *Mol. Cell. Biol.*, 12, 2919, 1992.

183. van Deursen, J., Wieringa, B., Targeting of the creatine kinase M gene in embryonic stem cells using isogenic and nonisogenic vectors, *Nucl. Acids Res.*, 20, 3815, 1992.

184. -te Riele, H., Maandag, E.R., Berns, A., Highly efficient targeting in embryonic stem cells through homologous recombination with isogenic DNA constructs, *Proc. Natl. Acad. Sci. U.S.A.*, 89, 5128, 1992.

185. Chang, X.B., Wilson, J.H., Modification of DNA ends can decrease end joining relative to homologous recombination in mammalian cells, *Proc. Natl. Acad. Sci. U.S.A.*, 84, 4959, 1987.

186. Berger, N.A., Poly(ADP-ribose) in the cellular response to DNA damage, *Radiat. Res.*, 101, 4, 1985.

187. Farzaneh, F., Panayotou, G.N., Bowler, L.D., Hardas, B.D., Broom, T., Walther, C., Shall, S., ADP-ribosylation is involved in the integration of foreign DNA into the mammalian celll genome, *Nucl. Acids Res.*, 16, 11319, 1988.

188. Waldman, B.C., Waldman, A.S., Illegitimate and homologous recombination in mammalian cells: differential sensitivity to an inhibitor of poly(ADP-ribosylation), *Nucl. Acids Res.*, 18, 5981, 1990.

189. Rossolini, G.M., Riccio, M.L., Gallo, E., Galeotti, C.L., Kluyveromyces lactis rDNA as a target for multiple integration by homologous recombination, *Gene*, 119, 75, 1992.

190. Ramdas, J., Mythili, E., Muniyappa, K., Nucleosomes on linear duplex DNA allow homologous pairing but prevent strand exchange promoted by RecA protein, *Proc. Natl. Acad. Sci. U.S.A.*, 88, 1344, 1991.

191. Wong, E.A., Capecchi, M.R., Homologous recombination between coinjected DNA sequences peaks in early to mid-S phase, *Mol. Cell. Biol.*, 7, 2294, 1987.

192. Bratthauer, G.L., Fanning, T.G., Active LINE-l retrotransposons in human testicular cancer, *Oncogene*, 7, 507, 1992.

193. Yeom, Y.I., Abe, K., Bennett, D., Artzt, K., Testis-/embryo expressed genes are clustered in the mouse H-2K region, *Proc. Natl. Acad. Sci. U.S.A.*, 89, 773, 1992.

194. Zimmer, A., Gruss, P., Production of chimaeric mice containing embryonic stem (ES) cells carrying a homeobox Hox 1.1 allele mutated by homologous recombination, *Nature*, 338, 150, 1989.

195. Koller, B.H., Smithies, O., Inactivating the beta 2-microglobulin locus in mouse embryonic stem cells by homologous recombination, *Proc. Natl. Acad. Sci. U.S.A.*, 86, 8932, 1989.

196. -te Riele, H., Maandag, E.R., Clarke, A., Hooper, M., Berns, A., Consecutive inactivation of both alleles of the pim-l proto-oncogene by homologous recombination in embryonic stem cells, *Nature*, 348, 649, 1990.

197. Oettinger, M.A., Activation of V(D)J recombination by RAGl and RAG2, *Trends Genet.*, 8, 413, 1992.

198. Gordon, J.W., Scangos, G.A., Plotkin, D.J., Barbosa, J.A., Ruddle, F.H., Genetic transformation of mouse embryos by microinjection of purified DNA, *Proc. Natl. Acad. Sci. U.S.A.*, 77, 7380, 1980.

199. Diacumakos, E.G., Methods for micromanipulation of human somatic cells in culture, *Methods Cell Biol.*, 7, 287, 1973.

200. Graham, F.L., van der Eb, A.J., A new technique for the assay of infectivity of human adenovirus 5 DNA, *Virology*, 52, 456, 1973.

201. Oi, V.T., Morrison, S.L., Herzenberg, L.A., Berg, P., Immunoglobulin gene expression in transformed lymphoid cells, *Proc. Natl. Acad. Sci. U.S.A.*, 80, 362, 1983.

202. Rice, D., Baltimore, D., Regulated expression of an immunoglobulin kappa gene introduced into a mouse lymphoid cell line, *Proc. Natl. Acad. Sci. U.S.A.*, 79, 7862, 1982.

203. Perucho, M., Hanahan, D., Wigler, M., Genetic and physical linkage of exogenous sequences in transformed cells, *Cell*, 22, 309, 1980.

204. Farber, F.E., Melnick, J.L., Butel, J.S., Optimal conditions for uptake of exogenous DNA by Chinese hamster lung cells deficient in hypoxanthine-guanine phosphoribosyltransferase, *Biochim. Biophys. Acta,* 390, 298, 1975.

205. Kucherlapati, R., Skoultchi, A.I , Introduction of purified genes into mammalian cells, *CRC Crit. Rev. Biochem.*, 16, 349, 1984.

206. McCutchan, J.H., Pagano, J.S., Enchancement of the infectivity of simian virus 40 deoxyribonucleic acid with diethylaminoethyl-dextran, *J. Natl. Cancer Inst.*, 41, 351, 1968.

207. Tur-Kaspa, R., Teicher, L., Levine, B.J., Skoultchi, A.I., Shafritz, D.A., Use of electroporation to introduce biologically active foreign genes into primary rat hepatocytes, *Mol. Cell. Biol.,* 6, 716, 1986.

208. Mukherjee, A.B., Orloff, S., Butler, J.D., Triche, T., Lalley, P., Schulman, J., Entrapment of metaphase chromosomes into phospholipid vesicles (lipochromosomes): carrier potential in gene transfer, *Proc. Natl. Acad. Sci. U.S.A.,* 75, 1361, 1978.

209. Bangham, A.D., Standish, M.M., Watkins, J.C., Diffusion of univalent ions across the lamellae of swollen phospholipids, *J. Mol. Biol.*, 13, 238, 1965.

210. Schaefer-Ridder, M., Wang, Y., Hofschneider, P.H., Liposomes as gene carriers: efficient transformation of mouse L cells by thymidine kinase gene, *Science*, 215, 166, 1982.

211. Slilaty, S.N., Aposhian, H.V., Gene transfer by polyoma-like particles assembled in a cell-free system, *Science*, 220, 725, 1983.

212. Bertling, W., Transfection of a DNA/protein complex into nuclei of mammalian cells using polyoma capsids and electroporation, *Biosci. Rep.*, 7, 107, 1987.

213. Vainstein, A., Razin, A., Graessmann, A., Loyter, A., Fusogenic reconstituted Sendai virus envelopes as a vehicle for introducing DNA into viable mammalian cells, *Methods Enzymol.*, 101, 492, 1983.

214. Loyter, A., Tomasi, M., Gitman, A.G., Etinger, L., Nussbaum, 0., The use of specific antibodies to mediate fusion between Sendai virus envelopes and living cells, *Ciba Found. Symp.*, 103, 163, 1984.

215. Bertling, W., Hunger-Bertling, K., Cline, M.J., Intranuclear uptake and persistence of biologically active DNA after electroporation of mammalian cells, *J. Biochem. Biophys. Methods*, 14, 223, 1987.

216. Calos, M.P., Lebkowski, J.S., Botchan, M.R., High mutation frequency in DNA transfected into mammalian cells, *Proc. Natl. Acad. Sci. U.S.A.*, 80, 3015, 1983.

217. Liljestrom, P., Garoff, H., A new generation of animal cell expression vectors based on the semliki forest virus replicon, *Bio/Technol.*, 1356, 1991.

218. Ishiura, M., Ohashi, H., Uchida, T., Okada, Y., Phage particle-mediated gene transfer of recombinant cosmids to cultured mammalian cells, *Gene*, 82, 281, 1989.

219. Kaneda, Y., Uchida, T., Kim, J., Ishiura, M., Okada, Y., The improved efficient method for introducing macromolecules into cells using HVJ (Sendai virus) liposomes with gangliosides, *Exp. Cell. Res.*, 173, 56, 1987.

220. Behr, J.P., Demeneix, B., Loeffler, J.P., Perez-Mutul, J., Efficient gene transfer into mammalian primary endocrine cells with lipopolyamine-coated DNA, *Proc. Natl. Acad. Sci. U.S.A.*, 86, 6982, 1989.

221. Cotten, M., Laengle-Rouault, F., Kirlappos, H., Wagner, E., Mechtler, K., Zenke, M., Beug, H., Birnstiel, M.L., Transferrin-polycation-mediated introduction of DNA into human leukemic cells: stimulation by agents that affect the survival of transfected DNA or modulate transferrin receptor levels, *Proc. Natl. Acad. Sci. U.S.A.*, 87, 4033, 1990.

222. Hooper, M., Hardy, K., Handyside, A., Hunter, S., Monk, M., HPRT-deficient (Lesch-Nyhan) mouse embryos derived from germline colonization by cultured cells, *Nature*, 326, 292, 1987.

223. van Deursen, J., Lovell-Badge, R., Oerlemans, F., Schepens, J., Wieringa, B., Modulation of gene activity by consecutive gene targeting of one creatine kinase M allele in mouse embryonic stem cells, *Nucl. Acids Res.*, 19, 2637, 1991.

224. Reid, L.H., Gregg, R.G., Smithies, O., Koller, B.H., Regulatory elements in the introns of the human HPRT gene are necessary for its expression in embryonic stem cells, *Proc. Natl. Acad. Sci. U.S.A.,* 87, 4299, 1990.

225. McClelland, A., Kamarck, M.E., Ruddle, F.H., Molecular cloning of receptor genes by transfection, *Methods Enzymol.*, 147, 280, 1987.

226. Charron, J., Malynn, B.A., Robertson, E.J., Goff, S.P., Alt, F.W., High-frequency disruption of the N-myc gene in embryonic stem and pre-B cell lines by homologous recombination, *Mol. Cell. Biol.*, 10, 1799, 1990.

227. Joyner, A.L., Skarnes, W.C., Rossant, J., Production of a mutation in mouse En-2 gene by homologous recombination in embryonic stem cells, *Nature*, 338, 153, 1989.

228. Jasin, M., Berg, P., Homologous integration in mammalian cells without target gene selection, *Genes Dev.*, 2, 1353, 1988.

229. Jasin, M., Elledge, S.J., Davis, R.W., Berg, P., Gene targeting at the human CD4 locus by epitope addition, *Genes Dev.*, 4, 157, 1990.

230. Le Mouellic, H., Lallemand, Y., Brulet, P., Targeted replacement of the homeobox gene Hox-3.1 by the *Escherichia coli* lacZ in mouse chimeric embryos, *Proc. Natl. Acad. Sci. U.S.A.*, 87, 4712, 1990.

231. Kim, H.S., Smithies, O., Recombinant fragment assay for gene targeting based on the polymerase chain reaction, *Nucl. Acids Res.*, 16, 8887, 1988.

232. Verma, I.M., Gene therapy, *Sci. Am.*, 263, 68, 1990.

233. Belmont, J.W., Caskey, C.T., Developments leading to gene therapy, in *Gene Transfer*, Kucherlapati, R., Ed., Plenum Press, New York, 1986, 411.

234. Kay, M.A., Ponder, K.P., Woo, S.L., Human gene therapy: present and future, *Breast Cancer Res. Treat.*, 21, 83, 1992.

235. Bradley, A., Evans, M., Kaufman, M.H., Robertson, E., Formation of germ-line chimaeras from embryo-derived teratocarcinoma cell lines, *Nature*, 309, 255, 1984.

236. Hogan, B., Costantini, F., Lacy, E., *Manipulating the Mouse Embryo. A Laboratory Manual*, Hogan, B., Costantini, F., and Lacy, E., Eds., Cold Spring Harbor Laboratory Press, Cold Spring Harbor, NY, 1986.

237. Stevens, L.C., Little, C.C., Spontaneous testicular teratomas in an inbred strain of mice, *Proc. Natl. Acad. Sci. U.S.A.*, 40, 1080, 1954.

238. Robertson, E., Bradley, A., Kuehn, M., Evans, M., Germ-line transmission of genes introduced into cultured pluripotential cells by retroviral vector, *Nature*, 323, 445, 1986.

239. Gossler, A., Doetschman, T., Korn, R., Serfling, E., Kemler, R., Transgenesis by means of blastocyst-derived embryonic stem cell lines, *Proc. Natl. Acad. Sci. U.S.A.*, 83, 9065, 1986.

240. Evans, M.J., Kaufman, M.H., Establishment in culture of pluripotential cells from mouse embryos, *Nature*, 292, 154, 1981.

241. Martin, G.R., Isolation of a pluripotent cell line from early mouse embryos cultured in medium conditioned by teratocarcinoma stem cells, *Proc. Natl. Acad. Sci. U.S.A.*, 78, 88, 1981.

242. Smith, A.G., Heath, J.K., Donaldson, D.D., Wong, G.G., Moreau, J., Stahl, M., Rogers, D., Inhibition of pluripotential embryonic stem cell differentiation by purified polypeptides, *Nature*, 336, 688, 1988.

243. Williams, R.L., Hilton, D.J., Pease, S., Willson, T.A., Stewart, C.L., Gearing, D.P., Wagner, E.F., Metcalf, D., Nicola, N.A., Gough, N.M., Myeloid leukaemia inhibitory factor maintains the developmental potential of embryonic stem cells, *Nature*, 336, 684, 1988.

244. Schwartzberg, P.L., Goff, S.P., Robertson, E.J., Germ-line transmission of a c-abl mutation produced by targeted gene disruption in ES cells, *Science*, 246, 799, 1989.

245. Mucenski, M.L., McLain, K., Kier, A.B., Swerdlow, S.H., Schreiner, C.M., Miller, T.A., Pietryga, D.W., Scott, W.J., Jr., Potter, S.S., A functional c-myb gene is required for normal murine fetal hepatic hematopoiesis, *Cell*, 65, 677, 1991.

246. Kitamura, D., Roes, J., Kuehn, R., Rajewsky, K., A B-cell deficient mouse by targeted disruption of the membrane exon of the immunoglobulin mu chain gene, *Nature*, 350, 423, 1991.

247. Koller, B.H., Marrack, P., Kappler, J.W., Smithies, O., Normal development of mice deficient in beta 2M, MHC class I proteins, and CD8+ T cells, *Science*, 248, 1227, 1990.

248. Zijlstra, M., Bix, M., Simister, N.E., Loring, J.M., Raulet, D.H., Jaenisch, R., Beta 2-microglobulin deficient mice lack CD4-8+ cytolytic T cells, *Nature*, 344, 742, 1990.

249. Blaese, R.M., Culver, K.W., Gene therapy for primary immunodeficiency disease, *Immunodefic. Rev.*, 3, 329, 1992.

250. Culliton, B.J., Designing cells to deliver drugs, *Science*, 246, 746, 1989.

251. Kasid, A., Morecki, S., Aebersold, P., Cornetta, K., Culver, K., Freeman, S., Lotze, M.T., Blaese, R.M., Anderson, W.F., Rosenberg, S.A., Human gene transfer: characterization of human tumor-infiltrating lymphocytes as vehicles for retroviral-mediated gene transfer in man, *Proc. Natl. Acad. Sci. U.S.A.*, 87, 473, 1990.

252. Sigal, S.H., Brill, S., Fiorino, A.S., Reid, L.M., The liver as a stem cell and lineage system, *Am. J. Physiol.*, 263, G139, 1992.

253. Matsumura, K., Lee, C.C., Caskey, C.T., Campbell, K.P., Restoration of dystrophin-associated proteins in skeletal muscle of mdx mice transgenic for dystrophin gene, *FEBS Lett.*, 320, 276, 1993.

254. Nicolau, C., Le Pape, A., Soriano, P., Fargette, F., Juhel, M.F., *In vivo* expression of rat insulin after intravenous administration of the liposome-entrapped gene for rat insulin I, *Proc. Natl. Acad. Sci. U.S.A.*, 80, 1068, 1983.

255. Stein, C.A., Cohen, J.S., Oligodeoxynucleotides as inhibitors of gene expression: a review, *Cancer Res.*, 48, 2659, 1988.

256. Knorre, D.G., Vlassov, V.V., Reactive oligonucleotide derivatives as gene-targeted biologically active compounds and affinity probes, *Genetica*, 85, 53, 1991.

257. Bertling, W.M., Aigner, T., Evidence for tissue specific activation of the retroposon L1 in mice, *J. Cell. Biochem.*, (suppl. 18B), 45, 1994.

258. Marschalek, R., personal communication, 1993.

Chapter 2

Molecular Mechanisms of Homologous Recombination

Alan S. Waldman

Department of Biological Sciences

University of South Carolina

Columbia, SC

Contents

I. Recombination basics

Homologous recombination (also sometimes called "general recombination") can be defined as any process in which similar DNA sequences exchange genetic information with one another. This type of recombination is distinguished from

0-8493-8950-X/95/$0.00+$.50

site-specific recombination or transpositional recombination in that, in homologous recombination, there is no requirement for a specific DNA sequence. Rather, homologous recombination may occur between any two sequences, provided that the two sequences share sufficient homology. With such a definition, one might imagine that there are a great many cellular processes that involve homologous recombination, and indeed, such is the case.

Gene targeting, or homologous recombination between a transfected DNA sequence and an homologous genomic sequence (the "target"), is an experimental application of a cell's ability to perform homologous recombination in order to bring about a desired alteration to an endogenous sequence. It has become common practice to use the term "homologous recombination" synonymously with "gene targeting." Although it is easy to understand why the power of gene targeting has stimulated great interest in homologous recombination among a wide array of investigators, this latter usage of vocabulary is not technically precise.

The aim of gene targeting is to alter a specific genomic sequence, usually for one of the following purposes: (1) learning about gene structure/function; (2) creating animal models for disease; or, perhaps, (3) ultimately curing genetic disease.[1,2] Accordingly, we may speak of two very broadly defined types of genetic alterations that are made possible by targeting strategies: restoration of genetic function or disruption of genetic function. Most current work involves the latter type of genomic alteration via the introduction of a mutation into a gene of interest. Such targeted alterations are commonly referred to as gene knockouts. Gene knockouts are usually intended to yield insight into the physiological role of a gene product.

The important natural roles that recombination plays in DNA repair and generation of genetic diversity have attracted many investigators over the past several decades to the study of recombination.[1,3-9] Many recombination studies have been carried out using fungal systems, largely because of the ease with which fungi can be grown, both mitotically and meiotically. Further, in filamentous fungi, all of the products of a single meiosis can be recovered in an ordered structure called the ascus, which has greatly facilitated the analysis of recombination events. From many decades of observations, some general notions about homologous recombination have emerged, and these concepts have applicability to our thinking about gene targeting.

There are two general ways in which DNA sequences can exchange information during homologous recombination: sequences can undergo either a reciprocal or a nonreciprocal exchange. As the terms suggest, a reciprocal exchange involves a swap of nucleotides between two sequences, whereas a nonreciprocal exchange involves a unidirectional exchange of nucleotides. These types of exchanges are often associated, that is, a nonreciprocal exchange may be coupled to a reciprocal exchange in an adjoining sequence. In a nonreciprocal exchange, it is possible to identify a donor and a recipient sequence with respect to the direction of flow of genetic information. A nonreciprocal exchange typically occurs with some loss of genetic information originally contained in the recipient, since part of the recipient sequence may be permanently altered by virtue of becoming identical to the donor sequence. Because of this loss of genetic information, nonreciprocal exchanges are sometimes referred to as nonconservative exchanges (as opposed to reciprocal exchanges, which are conservative). There is another way to define conservative vs. nonconservative recombination, which we will discuss later.

A nonreciprocal exchange is sometimes referred to as a gene conversion, since part of the recipient sequence is converted into a sequence identical to the donor. (To be precise, gene conversion was originally defined as a non-Medelian segregation of alleles manifested as a 3 to 1 segregation of alleles following meiosis in fungi.[10]) The stretch of DNA within the recipient sequence that has undergone conversion is referred to as the conversion tract.

A reciprocal exchange may be referred to as a crossover. It is possible for multiple crossovers to occur along a given segment of DNA. A single product of a double crossover event may resemble gene conversion, with the DNA segment between the two crossover points appearing as a conversion tract. However, if all of the products of a recombination event were to be examined, a double crossover would be distinguishable from a gene conversion due to the existence of a pair of reciprocal double crossover products. It is not always possible to recover all products of recombination, especially in mammalian systems, and therefore, in practice it is often not possible to distinguish with certainty between a double crossover and a gene conversion.[11]

The various modes of homologous recombination described above are reflected in two general classes of gene targeting events, which are often described as gene replacement vs. targeted insertion.[2] In gene replacement, an endogenous sequence is precisely replaced with a homologous sequence contained on the targeting vector (see Figure 1a). In a gene replacement, there need not be a change in the number of nucleotides in the genome. In targeted insertion, the entire transfected molecule, including the vector sequences, becomes inserted at the target locus. After targeted insertion, two copies of a target sequence reside in the genome flanking an integrated copy of the vector (Figure 1b). For most applications, gene replacement would likely be preferred over insertion. This is because gene replacement need not augment the genome with foreign sequences (i.e., vector sequences). Also, the closely linked duplicated target sequences resulting from targeted insertion are genetically unstable since the repeated sequences may undergo intrachromosomal recombination with one another.

It has been shown[12] that gene replacements can be promoted by cleaving the targeting construct within vector sequences prior to transfection so that the targeted homology on the transfected molecule is colinear with the genomic target. Targeted insertions can be promoted by cleaving the transfected molecule within the targeted homology to produce adjacent DNA termini within homology. Depending on the manner in which the transfected molecule is cut, the transfected molecule is referred to as either a gene replacement construct or an insertion construct.

Either a gene conversion or a double reciprocal exchange between target and transfected sequences can give rise to a gene replacement. As discussed

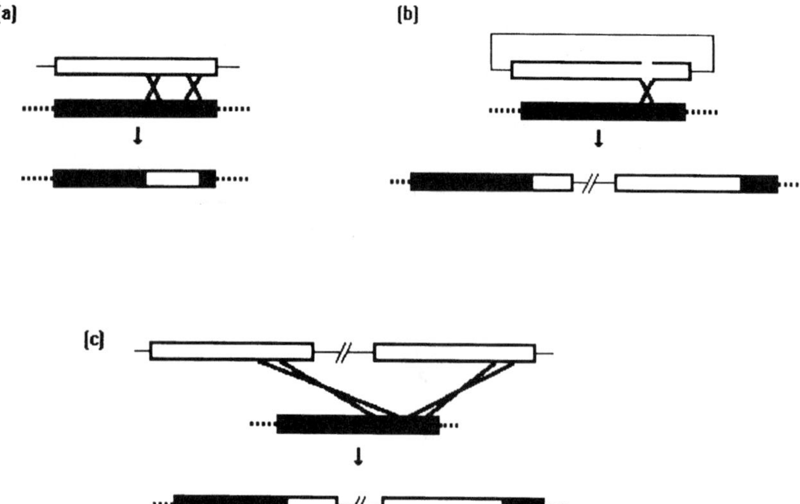

Figure 1. Modes of gene targeting. In this representation of gene targeting events, the target sequence is denoted by the black rectangle, with flanking genomic sequences denoted by the broken line. The targeted homology contained on the transfected construct is represented by the white rectangle, with vector sequences depicted by thin lines. (a) Gene replacement by a construct cleaved within vector sequences. Part of the genomic target is replaced by the transfected sequence, while the fate of the transfected construct is usually unknown. Gene replacement can occur by a double crossover or by gene conversion. (b) Targeted insertion by a construct cleaved within targeted homology. Insertion results in two copies of the target sequence flanking an integrated copy of the vector. Insertion may occur by a single crossover at the break in homology, as shown, or by a single crossover involving a circular construct. (c) Apparent targeted insertion via gene replacement involving a concatemer of the transfected construct. As illustrated, such a gene replacement may yield a product that is indistinguishable from a targeted insertion.

above, these two types of events can be distinguished only by recovering all recombination products. Since this is not possible during gene targeting, that is, the fate of the participating transfected molecule is usually unknown, the predominant mechanism of gene replacement remains an unknown.

There are actually a few reports in which the fate of the recombining targeting construct was determined. In some early studies of gene targeting in mammalian cells, the experimental strategy involved the reconstruction of a selectable genetic marker so that the gene targeting events were directly selectable.[2] In such a scenario, there are theoretically two ways in which the marker can be restored: correction of the target or correction of the defective gene on the transfected molecule followed by random integration of the corrected gene. Both types of occurrences have been reported.[2,13-16] In the cases reported in which a transfected construct was corrected by the genomic target,

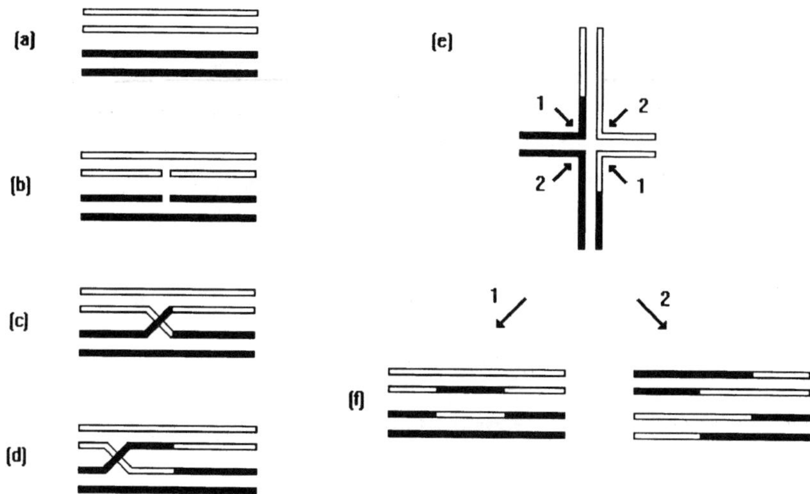

Figure 2. *Schematic illustration of the Holliday model[18] for homologous recombination. (**a**) Homologous pairing of duplexes. (**b**) Strand breakage at homologous positions in strands of like polarity. (**c**) Strand invasion and sealing of breaks, establishing the Holliday structure. (**d**) Branch migration of crossover point resulting in elongation of heteroduplex. (**e**) Alternate depiction of Holliday structure after branch migration. Two alternate modes of resolution of Holliday junction are shown. Resolution involves a "diagonal" cleavage through the central four-way junction. (**f**) Resolution pathway 1 leads to noncrossover products; pathway 2 leads to crossover products. Please see text for details.*

there was no evidence that the genomic target had been altered, suggesting the likelihood that in these cases a gene conversion mechanism was involved.

(It is not presently clear what factors are important in directing the flow of genetic information during gene targeting. Resolving this issue is of clear practical concern. Thomas et al.[13] suggested that mammalian cells might have a propensity to correct deletions rather than insertions, that is, it might be wiser to try to introduce an insertion rather than a deletion mutation into a genomic target so that information flows to, rather than from, the genome. This generalization has not been confirmed.)

Like gene replacement, targeted insertion can arise from several different mechanisms. Crossing-over initiated from a break within the targeted homology on a transfected molecule would yield an insertion (Figure 1b), as would a single crossover between a circular transfected molecule and the target. Certain gene conversions or double crossovers between a concatemeric transfected molecule and the target can also yield apparent insertion products (Figure 1c). It should be clear from this discussion that the definitions of the terms "replacement" and "insertion" are phenomenological rather than mechanistic, and the distinction between the two types of events is perhaps at times artificial.

II. Recombination models

How does a living cell carry out homologous recombination? This is a question that has challenged investigators for many years. As explained above, fungal systems are particularly amenable to the study of recombination, and indeed, many of the present day recombination models are based on studies conducted in yeast and other fungi. The reader is encouraged to refer to an excellent review on fungal recombination[9] for details of the models that will receive a brief treatment here.

Probably the biggest breakthrough in conceptualizing the manner in which homologous sequences may exchange genetic information came in the mid-1960s with publications by Whitehouse[17] and Holliday.[18] In these seminal works, the "copy-choice hypothesis", which envisioned recombination as occurring as a consequence of template switching during replication of chromosomes,[17] was effectively laid to rest. The works by Whitehouse and Holliday largely dealt with hybrid deoxyribonucleic acid, now commonly referred to as heteroduplex DNA. The term "heteroduplex DNA" refers to the establishment of an interval of double-stranded DNA during recombination in which one strand is contributed by one of the recombining parental DNA molecules while the complementary strand is contributed by the other parental DNA molecule. Heteroduplex DNA provided a plausible scenario for how two homologous sequences could precisely swap information, since the problem of establishing the required perfect alignment between recombining sequences was solved through base-pairing. In a region of heteroduplex DNA, there are likely to be several mismatched, noncomplementary bases that the cell may repair in order to restore complementarity. The repair of heteroduplex DNA back to homoduplex DNA was viewed by Whitehouse[17] and Holliday[18] as the mechanism for gene conversions.

Heteroduplex repair is still viewed as a major mechanism of gene conversion. In the simplest of cases, such repair can use either strand of the heteroduplex as template, so that there would be a 50% chance that repair of a segment of heteroduplex produces a gene conversion and a 50% chance that the heteroduplex correction restores the original sequences. There is the additional possibility of patchy or discontinuous repair, and there has been some evidence for this phenomenon in gene targeting.[19]

Holliday's model for recombination[18] was the first model to become widely accepted. In a conceptually simple way, the Holliday model presented a plausible way to envision recombination and explained both the observed association between reciprocal exchanges and gene conversions as well as the occurrence of gene conversions in the absence of crossing-over. The Holliday model, the basic steps of which are presented in Figure 2, presented a paradigm upon which most later models have been based and is, therefore, worthy of some further discussion.

An initial step in the Holliday model (and virtually all subsequent models) is homologous pairing of the homologous sequences (Figure 2a). (How this is accomplished was not and is presently not entirely understood.) Once strong interactions are established between the paired sequences, one may say that synapsis has occurred, and the tightly paired sequences along with associated pairing proteins are collectively referred to as a synaptic complex.

Sometime after pairing, strand breakage occurs (Figure 2b). (The details of how strand breakage occurs have subsequently been revised in newer recombination models, but, in general, if two sequences are going to exchange nucleotides, one or more strands must break during the process.) In the Holliday model, single-strand breakage occurs in the paired sequences at homologous locations in strands of like polarity. Strand invasion then occurs at the break sites as the broken strands from each duplex simultaneously swap places with one another and invade the other duplex (Figure 2c). In this process, heteroduplex DNA is formed symmetrically. After strand invasion, the strand breaks are sealed and the two duplexes are covalently joined to each other at a point of single-strand crossover (Figure 2c). Holliday envisioned that this crossover point could spontaneously migrate, and this process is referred to as branch migration. In theory, branch migration may be accomplished at no expense of energy because as one basepair is melted, another is immediately formed. As branch migration proceeds, the stretch of heteroduplex is lengthened (Figure 2d).

A key feature of the Holliday model is the so-called Holliday structure, which is still viewed as the central recombination intermediate in most recombination paradigms. The Holliday structure is formed once the paired sequences are covalently linked and strand breaks are sealed (as in Figure 2c-d). The structure is essentially defined by the point of strand crossover, as well as the sequences in close proximity to this point. A Holliday structure can be viewed as being equivalent to a branched DNA structure, with four duplex arms containing a central four-way junction (the Holliday junction) as depicted in Figure 2e. The terminal steps of recombination involve cleavage, or resolution, of the Holliday structure through the central junction. The Holliday junction is viewed as being essentially symmetrical and can be resolved in one of two ways (as shown in Figure 2e). One mode of cleavage yields a noncrossover product (cleavage pathway 1, Figure 2f), while the other cleavage yields a product with a crossover (cleavage pathway 2, Figure 2f). Regardless of the mode of resolution, the establishment of heteroduplex DNA during strand invasion and branch migration (and the subsequent repair of the heteroduplex) allows gene conversion to occur in either case, as described above.

In 1979, Potter and Dressler were able to directly visualize the formation of Holliday stuctures in *E. coli* using electron microscopy,[20] providing some corroboration of the Holliday model. Additonally, if the various steps of recombination portrayed by the Holliday model are valid, one would predict that there must exist in living cells a machinery capable of carrying out the various steps. Quite a number of proteins that can catalyze various aspects of the general process depicted by Holliday have, in fact, been identified in various organisms (ranging from bacteria to humans). A discussion of what is known about the biochemistry of recombination is beyond the scope of this chapter, and the reader is referred to recent reviews in this area.[3,4]

Most of the features of the Holliday model have survived in the myriad models that have appeared subsequently, although some of the finer details have been revised. An important revision was made by Meselson and Radding in 1975,[21] in response to the observation that recombination does not usually occur symmetrically. As described above, the original Holliday model predicted the formation of symmetric heteroduplex due to simultaneous strand invasion of both duplexes in the synaptic complex. Meselson and Radding proposed that strand invasion may be initiated assymetrically. In the newer model, a strand from one duplex may initially invade the other duplex with no reciprocal strand invasion. The strand invasion is initiated by a single-strand break on the invading strand. DNA synthesis initiating at the break site drives the invasion by displacing the invading strand. In this fashion, the Meselson/Radding model links recombination to DNA synthesis. This is a departure from the Holliday model, in which no DNA replication is required for recombination to occur. As strand invasion proceeds, the displaced strand of the invaded duplex forms an increasingly large D-loop. Eventually, the D-loop breaks and a free end of DNA from the broken D-loop invades the other duplex. Breaks in strands are eventually sealed, and the two duplexes are covalently joined in a Holliday structure. From this point on, the Meselson/Radding model is identical to the Holliday model.

A different model, called the double-strand break-repair model (DSBR) (Figure 3), was later proposed[22] in order to account for the observation that double-strand breaks greatly increase recombination rates in yeast. Neither the Meselson/Radding model nor the Holliday model provided a plausible explanation for such an observation. The stimulatory effect of double-strand breaks has been often observed in mammalian cells as well.[1]

Essentially, the DSBR model envisions that recombination is initiated at the site of a double-strand break, which is enlarged into a double-strand gap (Figure 3 a-c). The gap is repaired using as template a strand of DNA from an homologous duplex to which the broken sequence is paired (Figure 3d-f). The repaired gap defines a conversion tract. Gene conversion in this model may, therefore, be accomplished in the absence of heteroduplex repair (although short stretches of heteroduplex DNA may flank the gap). Another essential difference between the DSBR and Meselson/Radding models is that the broken sequence in the DSBR model acts primarily as a recipient of genetic information, whereas in the Meselson/Radding model the broken sequence acts as a genetic donor.

During the course of DSBR, a double Holliday structure is formed, with the two Holliday junctions flanking the conversion tract (Figure 3f). Each Holliday junction can be resolved in two fashions, for a total of four possible modes of resolution. Two of these resolution modes yield crossover products, whereas the other two resolution modes yield noncrossover products (Figure 3g). The DSBR model, therefore, incorporates some of the important concepts originally introduced by the Holliday model. The DSBR model has recently received minor revision to accommodate certain recombination events in yeast.[23]

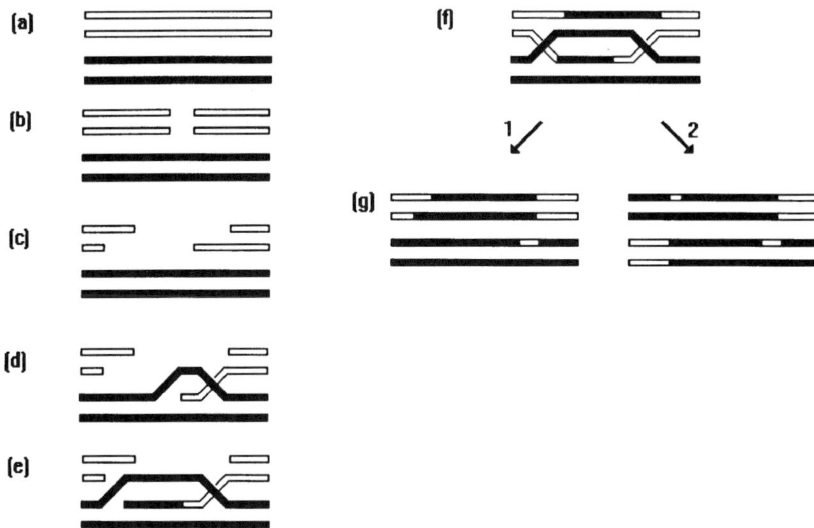

*Figure 3. Schematic illustration of the double-strand break-repair (DSBR) model[22] for homologous recombination. (**a**) Homologous pairing of duplexes. (**b**) A double-strand break is introduced in one duplex. (**c**) The break is enlarged into a double-strand gap. (**d**) Strand from broken sequence invades unbroken duplex, forming a D-loop. (**e**) D-loop is enlarged due to replication primed from invading strand. D-loop "invades" gapped duplex and establishes a short segment of heteroduplex at left end of gap. (**f**) Gap is repaired using strand from D-loop as a template. Ligation of DNA termini forms a double Holliday structure. (**g**) The double Holliday structure is resolved in one of four possible ways, two of which are shown. Resolution pathway 1 yields noncrossover products; pathway 2 yields crossover products.*

A recombination model that is very different from the Holliday, Meselson/Radding, or DSBR models was developed by Lin et al.[24] to explain some observations of both positive and negative effects of double-strand breaks on extrachromosomal recombination rates in mammalian cells. The model has become known as single-strand annealing (SSA) (Figure 4). In SSA, recombination initiates at the site of a double-strand break by bidirectional single-strand-specific exonucleolytic digestion (Figure 4a,b). This degradation continues until two complementary single-stranded sequences are exposed. The complementary strands then anneal (Figure 4c). Single-stranded unannealed sequence tails are clipped, and any single-strand sequence gaps are repaired (Figure 4d).[24]

There are at least four important differences between SSA and the Holliday, Meselson/Radding, and DSBR models. In SSA, (1) there is no step that involves pairing of DNA duplexes, and so there is no true homology search involved; (2) no Holliday structures are formed as intermediates; (3) recombination products are exclusively crossovers; and (4) recombination is

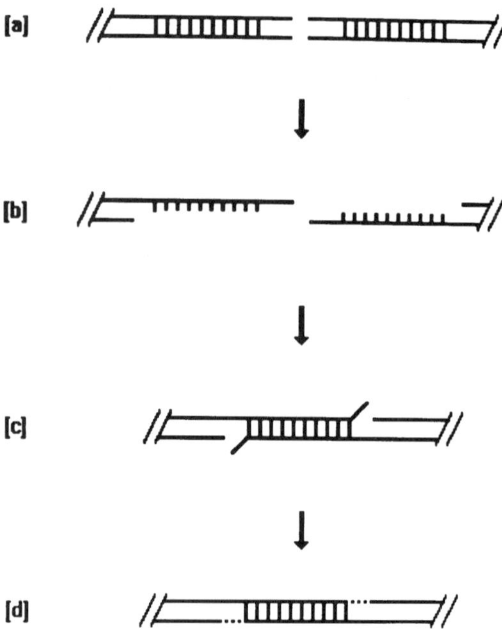

*Figure 4. The essential steps of the single-strand annealing (SSA) model[24] for nonconservative recombination. (**a**) In this simplified schematic, two homologous sequences are depicted by the parallel vertical lines. A double-strand break is positioned between the two homologous sequences. (The two homologous sequences need not be genetically linked to one another, that is, they may reside on two different DNA molecules.) (**b**) Single-strand degradation proceeds from the DNA break site, exposing complementary single-stranded sequences. (**c**) The complementary single strands anneal to one another. (**d**) Unannealed DNA tails are clipped and single-strand gaps are repaired. Note that the SSA model starts with two homologous sequences and ends with only one copy of the sequence.*

nonconservative, that is, recombination necessarily proceeds with the net loss of one of the recombining sequences due to the degradative mechanism (Figure 4). (The recombination product is constructed from the Watson strand of one parental homologous sequence and the Crick strand of the other parent. The Crick strand of the first parental sequence and the Watson strand of the second parent are destroyed during SSA.)

III. Gene targeting mechanisms

What is the mechanism of gene targeting in mammalian cells? Much of our insight into this difficult issue derives from studies of both extrachromosomal and intrachromosomal recombination in mammalian cells. One should appreciate that several differences between these latter two modes of recombination have been recorded, which begs the question of whether gene targeting more closely resembles extra- or intrachromosomal recombination. Perhaps the two most striking (or at least the most reported) differences between extra- and intrachromosomal recombination in mammalian cells are the homology requirements of the recombination events and the issue of whether recombination is conservative vs. nonconservative (in the sense defined by SSA).

Intrachromosomal recombination has been found to be a great deal more sensitive to small degrees of heterology than is extrachromosomal recombination.[1,25] Waldman and Liskay[26] reported that about 150 bp of contiguous homology are required for efficient intrachromosomal recombination in mouse fibroblasts, and that perhaps even a single nucleotide mismatch embedded in a 200 bp stretch of homology can reduce intrachromosomal recombination rate by more than an order of magnitude. Work by Deng and Capecchi[27] indicated that such exquisite sensitivity to heterology is not the rule for gene targeting in mouse embryonic stem (ES) cells. Targeting between sequences from nonisogenic strains of mice was generally found to be only a few fold less efficient than gene targeting between sequences from isogenic strains.[27]

Intrachromosomal recombination in mouse cells was shown to be inversely proportional to the length of a heterologous sequence embedded within one of the two recombining sequences.[28] No such sensitivity was observed in gene targeting efficiency in experiments involving the transfer of embedded heterologies of up to 12 kb.[29] From the standpoint of sensitivity to heterology, it might, therefore, be concluded that gene targeting does not greatly resemble intrachromosomal recombination.

Gene targeting, like both intra- and extrachromosomal recombination, is apparently sensitive to the length of homology shared by the recombining sequences.[12,30,31] In at least one report,[12] the rate of targeting was essentially exponentially sensitive to homology length. It has been recommended[32] that at least 1 kb of targeted homology flank any embedded heterology residing in a gene replacement type of construct. This is to ensure that gene targeting is not only efficient but accurate as well, since limited homology has been associated with aberrant gene targeting events when gene replacement constructs are used.[32]

In recent years, the notion that virtually all extrachromosomal homologous recombination in mammalian cells proceeds in a nonconservative fashion (via SSA), whereas intrachromosomal recombination proceeds conservatively (with no loss of a recombining sequence), has gained fairly wide acceptance and has prompted at least one group to directly assess whether gene targeting is conservative or nonconservative. The study conducted by Pennington and Wilson[33] indicated that targeting in Chinese hamster ovary (CHO) cells is, in fact, conservative and, therefore, is not mediated by SSA. (It should be noted that recent work by Yang and Waldman[34] challenges the notion that all, or even most, extrachromosomal recombination in mammalian cells, in fact, proceeds via SSA.)

The work by Pennington and Wilson[33] (that showed that gene targeting is conservative) was performed with an insertion type of construct, that is, the targeting construct was cleaved within targeted homology (in this case an aprt gene sequence). In this work, it was observed that genetic markers in the vicinity of the double-strand break were replaced by markers from the genomic target. This result was consistent with a DSBR paradigm (Figure 3), in which the DNA break had been enlarged into a gap and was then repaired using information from a homologous sequence (that is, the target). Additional recent evidence indicates that DSBR may be a major mechanism for gene targeting in mammalian cells.[35] This evidence was largely based on the observation that transfected molecules containing large gaps within the targeted homology were capable of correcting a defective hprt gene in mouse ES cells and, further, that the gaps in the transfected sequence were repaired using genetic information contributed by the genomic target sequences. Additional recent studies using mouse ES cells[36] also revealed results consistent with DSBR.

As discussed above, the development of the DSBR model for recombination was originally prompted by the observation that a double-strand break within homology can stimulate recombination. If gene targeting in mammalian cells truly proceeds via DSBR, then one might expect that a double-strand break within homology would stimulate gene targeting. This prediction was borne out in several studies.[2] In other studies, however, double-strand breaks within targeted homology were no more effective at stimulating targeting than were breaks placed within vector sequences on the targeting vehicle.[2,52] It is not clear what to make of these different results in terms of the DSBR paradigm; further investigation is needed.

The entire issue of the role of DNA ends in the gene targeting process has stirred some controversy in the literature. Some investigators[27,32] have observed that there is no direct participation of free DNA ends in the gene targeting mechanism. This observation, in fact, played a pivotal role in the development of the successful and widely used positive-negative selection scheme for enriching for gene targeting events.[37] In this scheme, the concept that random integration proceeds via DNA ends, whereas targeting does not, is exploited to allow a negative selection against random uptake of a herpes thymidine kinase gene placed at the terminus of a gene replacement construct. The group that developed the positive-negative selection scheme also reported that gene replacement constructs (broken within vector sequence) perform targeting in ES cells as efficiently as do insertion constructs (broken within targeted homology).[12]

Other investigators[38,39] have reported that a double-strand break placed within the targeted homology is crucial for efficient targeting in mouse ES cells. It has been reported that gene replacement vectors are intrinsically inefficient and their use leads to many aberrant events.[38] Based on such observations, a model for gene targeting has been proposed by Hasty et al.,[39] in which adjacent DNA ends within homology play a critical role in the targeting process in a variation of the DSBR paradigm. In this targeting model, two adjacent free DNA ends from the same broken strand of DNA from an insertion construct simultaneously invade the target sequence, forming a single D-loop. The ends of the invading strands often ligate within the D-loop. The event is then processed as in the DSBR model. In this paradigm, the inefficiency of targeting using gene replacement constructs is seen as stemming from the need for such events to proceed via the formation of two D-loops. However, it should be noted that, as described by the Holliday and Meselson/Radding models discussed above, gene conversions (which may be responsible for gene replacements) can proceed via the formation of only a single D-loop. Also, it has been observed[1] that mammalian cells are quite capable of performing gene conversion-like events and may actually carry out conversions more frequently than single crossovers.

In the work leading to the model by Hasty et al.,[39] the vast majority of gene targeting events recovered were resolved as insertion events, although there is no apparent reason, a priori, why this should have been the case based on the model. The model does predict that transfected molecules cut within targeted homology should work more efficiently than ones cut within vector sequences, but the model does not explain why events should be resolved in favor of insertion over replacement. In other works involving CHO cells, a targeting construct cut within targeted homology and, therefore, presumably primed for insertion, actually gave rise to a majority of gene replacement events.[40,52]

From this discussion, it should be clear that the mechanism(s) of gene targeting in mammalian cells is far from being solved. The apparently conflicting results obtained may be a reflection of either multiple targeting mechanisms or complex effects of specific sequences, cell types, and/or genetic selections on the targeting process. One should also keep in mind that gene conversion events involving concatemers may look like insertion events, so the distinction between insertion and replacement is not always clear.

IV. Thoughts on how to optimize the efficiency of gene targeting in mammalian cells

To this point, this chapter has dealt with mechanisms of homologous recombination and possible mechanisms for gene targeting, but a central issue regarding gene targeting in mammalian cells has yet to be discussed. A major obstacle that stands in the way of applying gene targeting to such useful ends as gene therapy is the inefficiency of the gene targeting process in mammalian cells. Roughly speaking, about one in 10^6 mammalian cells transfected will successfully perform gene targeting, and the ratio of random integration to gene targeting in transfected cells is typically about[2] 1000 to 1. How may these odds be overcome? One approach is to use clever genetic tricks to be able to identify and isolate the rare targeting events. These tricks have been employed quite productively, especially in producing gene knockouts.[2] Another approach is to increase the absolute efficiency of gene targeting. An absolute increase in efficiency would ostensibly be mandatory for gene targeting to be truly useful for gene therapy.

How can we improve the efficiency of gene targeting if the mechanism is so poorly understood? Although we do not know the details of how targeting is accomplished, sufficient insight into recombination has been gained over the past decades to provide us with some rational approaches as starting points. As mentioned above, it is known that all types of homologous recombination events that have been investigated in mammalian cells are sensitive to the length of shared homology between the recombining sequences. This knowledge has already been put to use to improve targeting efficiency and accuracy.[32] Perhaps if exceptionally long homology (say >50 kb) were used in a targeting vector, the efficiency would be improved dramatically. This is a possibility through the use of yeast artificial chromosomes ("YACS").

There is some evidence that levels of homologous recombination in mitotically dividing cells is cell cycle dependent.[1] It is conceivable that the transfection of synchronized cell cultures might be useful to attempt to hit upon the proper window of opportunity for gene targeting, if indeed such a window exists. It is also conceivable that certain culture conditions, such as serum deprivation, for example, might stimulate gene targeting by an as of yet unknown mechanism.

The precise mode of transfection might also influence targeting efficiency. There is some evidence, for example, that nuclear microinjection yields a higher targeting efficiency than does either calcium phosphate/DNA coprecipitation or electroporation.[2] (It is, however, more difficult to transfect large numbers of cells by microinjection than with the latter two methodologies.) Development of new transfection methodologies may, therefore, have some merit.

One aspect of gene targeting that has received much attention is the issue of the influence of DNA breaks on recombination rate, as described above. As the literature currently stands, it would seem wise to use linearized DNA in gene targeting attempts. Although the precise role of DNA termini is not clear, it is a common observation that linear molecules undergo targeting at a greater efficiency than do circular molecules.[2] Linear DNA molecules may be topologically better suited for gene targeting than are circular molecules. It is not clear that breaking the transfected molecule within targeted homology is particularly useful.

Perhaps one relevant aspect of the effect of DNA breaks on extrachromosomal recombination in mammalian cells is the observation that a broken molecule tends to act as an efficient recipient of genetic information. This is consistent with and predicted by the original DSBR model.[22] To some extent, this observation has been corroborated in gene targeting events.[35] The problem is that, whereas it is very easy for an investigator to place a break within the transfected sequence (which many investigators have indeed done), it is the genomic sequence that is intended to be the recipient of genetic information in the targeting transaction, and so it is the genomic target that perhaps ought to be broken. Currently, methodology does not exist to accomplish this task.

The real issue at the heart of the matter of improving gene targeting efficiency is the question of what constitutes the rate-limiting step of gene targeting, since it is the rate-limiting step that must be overcome. Several reports in the literature suggest that the initial interaction between a transfected molecule and a genomic target is not rate limiting. This conclusion is based on reports that neither increasing the amount of transfected DNA[13] nor increasing the number of copies of the target[41] improves targeting efficiency. (If this is truly the case, then the inefficiency of gene targeting in mammalian cells compared to yeast is not due to the dilution of the target in a larger genome.) These reports should, perhaps, be interpreted with caution, since extrachromosomal recombination among transfected sequences may compete with targeting and, hence, offset any benefit an increased amount of transfected DNA has on gene targeting. Additionally, in the study in which the copy number of the target was altered,[41] most copies of the target were in a tandem array. Homologous sequences arranged in tandem may engage in interactions that may interfere with targeting, and so the behavior of a tandem array may be very different from that of dispersed targets.

The notion that gene targeting is independent of the concentration of either of the two recombining sequences is rather counterintuitive and difficult to fathom. If this notion were correct, it would be inappropriate to focus our attention on the interacting DNA sequences in our attempts to improve targeting efficency. However, it is probably wise to consider the question of the rate-limiting step as still open for debate.

If we accept the possibility that the slow step of targeting may very well be the initial interaction between the recombining sequences, then approaches toward blocking or inhibiting random integration as a means to improve targeting seem reasonable. If random integration is inhibited, then a greater portion of the transfected DNA molecules may be channeled toward gene targeting. The finding of randomly integrated copies of transfected DNA within the genomes of some cells that have undergone gene targeting[2] suggests that there is validity to investigating ways of blocking random integration.

How may we block random integration? Farzaneh et al.[42] reported that inhibition of poly(ADP-ribosylation) during transfection of mammalian cells inhibits random integration of DNA (possibly by inhibiting DNA end-joining processes). Subsequently, it has been shown that inhibition of poly(ADP-ribosylation) by 3-methoxybenzamide does not inhibit extrachromosomal homologous recombination[43] and actually mildly stimulates intrachromosomal recombination[44] in mouse cells. These works suggest that inhibition of poly(ADP-ribosylation) may be experimentally useful at improving gene targeting efficiency, or at least at reducing the background of unwanted random integrations. The impact of such an approach on gene targeting is yet to be seen.

Other possible approaches to improve targeting efficiency include transfecting cells with chromatin rather than naked DNA, since the manner in which mammalian cells handle transfected chromatin has not been thoroughly explored. Another possibility is to take advantage of the observation that transcription enhances recombination in yeast and mammalian cells.[45-47] Although active transcription of the target sequence seems not to significantly influence targeting efficiency,[2] there is evidence that active transcription of the transfected sequence might.[48] Perhaps a strong promoter on the transfected molecule would be helpful.

One may look at the inefficiency of gene targeting from a different standpoint and recognize that gene targeting is in effect no less efficient than intrachromosomal homologous recombination. Spontaneous recombination between two closely linked repeated sequences occurs only about once per 10^6 generations in mitotically dividing cells.[1] For this reason, perhaps we should expect gene targeting to occur in no more than about one in 10^6 transfected cells, since the window of opportunity for gene targeting in a typical targeting experiment is only one or very few generations following transfection. Perhaps, therefore, gene targeting that is described in many published reports is indeed occurring at maximal possible spontaneous rates, and the only way to improve the efficiency would be to artificially induce high levels of homologous recombination in the cells. This could, in theory, be accomplished by subjecting cells to certain conditions known to stimulate recombination activity, such as exposure of cells to mutagens[49] or subjecting cells to thymidylate stress.[50] The problem is that such treatments may induce genomic mutations, and also, there has been no reported cellular treatment that increases recombination more than a fewfold.

Finally, another potential approach to induce high levels of gene targeting might be to introduce a limiting recombination protein(s) into the cells along with the transfected DNA. In principle, this protein need not be from a mammalian source. Perhaps prior to transfection, the "recombinase" could be assembled onto the transfected molecule *in vitro*, in a fashion similar to the manner in which the recA protein of *E. coli* forms a nucleoprotein filament.[51] In the absence of further studies into the recombination mechanisms in higher eukaryotes, it is difficult to make an assessment of the feasibility of such an approach.

References

1. Bollag, R.J., Waldman, A.S., Liskay, R.M., Homologous recombination in mammalian cells, *Annu. Rev. Genet.*, 23, 199, 1989.

2. Waldman, A.S., Targeted homologous recombination in mammalian cells, *Crit. Rev. Oncol. Hematol.*, 12, 49, 1992.

3. Cox, M.M., Lehman, I.R., Enzymes of general recombination, *Annu. Rev. Biochem.*, 56, 229, 1987.

4. West, S.C., Enzymes and molecular mechanisms of genetic recombination, *Annu. Rev. Biochem.*, 61, 603, 1992.

5. Kucherlapati, R., Moore, P.D., Biochemical aspects of homologous recombination in mammalian somatic cells, in *Genetic Recombination*, Kucherlapati, R. and Smith, G.R., Eds., American Society for Microbiology, Washington, DC, 1988, 575.

6. Kucherlapati, R., Homologous recombination in mammalian somatic cells, in *Gene Transfer*, Kucherlapati, R., Ed., Plenum Press, New York, 1986, 363.

7. Subramani, S., Seaton, B.L., Homologous recombination in mitotically dividing mammalian cells, in *Genetic Recombination*, Kucherlapati, R. and Smith, G.R., Eds., American Society for Microbiology, Washington, DC, 1988, 549.

8. Würgler, F.E., Recombination and gene conversion, *Mutat. Res.*, 284, 3, 1992.

9. Orr-Weaver, T.L., Szostak, J.W., Fungal recombination, *Microbiol. Rev.*, 49, 33, 1985.

10. Zickler, H., Genetische untersuchungen an einem heterothallischen askomyzeten (Bombardia lunata nov. spec.), *Planta*, 22, 573, 1934.

11. Liskay, R.M., Stachelek, J.L., Evidence for intrachromosomal gene conversion in cultured mouse cells, *Cell,* 35, 157, 1983.

12. Thomas, K.R., Capecchi, M.R., Site-directed mutagenesis by gene targeting in mouse embryo-derived stem cells, *Cell*, 51, 503, 1987.

13. Thomas, K.R., Folger, K.R., Capecchi, M.R., High frequency targeting of genes to specific sites in the mammalian genome, *Cell*, 44, 419, 1986.

14. Jasin, M., de Villiers, J., Weber, F., Schaffner, W., High frequency of homologous recombination in mammalian cells between endogenous and introduced SV40 genomes, *Cell*, 43, 695, 1985.

15. Shaul, Y., Laub, O., Walker, M.D., Rutter, W.J., Homologous recombination between a defective virus and a chromosomal sequence in mammalian cells, *Proc. Natl. Acad. Sci. U.S.A.,* 82, 3781, 1985.

16. Subramani, S., Rescue of chromosomal T-antigen sequences onto extrachromosomally replicating, defective simian virus 40 DNA by homologous recombination, *Mol. Cell. Biol.*, 6, 1320, 1986.

17. Whitehouse, H.L.K., A theory of crossing-over by means of hybrid deoxyribonucleic acid, *Nature*, 199, 1034, 1963.

18. Holliday, R., A mechanism for gene conversion in fungi, *Genet. Res. Camb.*, 5, 282, 1964.

19. Brinster, R.L., Braun, R.E., Lo, D., Avarbok, M.R., Oram, F., Palmiter, R.D., Targeted correction of a major histocompatibility class II Ea gene by DNA microinjected into mouse eggs, *Proc. Natl. Acad. Sci. U.S.A.*, 86, 7087, 1989.

20. Potter, H. Dressler, D., On the mechanism of genetic recombination: electron microscopic observation of recombination intermediates, *Proc. Natl. Acad. Sci. U.S.A.*, 73, 3000, 1976.

21. Meselson, M.S., Radding, C.M., A general model for genetic recombination, *Proc. Natl. Acad. Sci. U.S.A.*, 72, 358, 1975.

22. Szostak, J.W., Orr-Weaver, T.L., Rothstein, R.J., Stahl, F.W., The double-strand break-repair model for recombination, *Cell*, 33, 25, 1983.

23. Song, H., Treco, D., Szostak, J.W., Extensive 3'-overhanging, single-stranded DNA associated with the meiosis-specific double-strand breaks at the ARG4 recombination initiation site, *Cell*, 64, 1155, 1991.

24. Lin, F.-L., Sperle, K., Sternberg, N., Model for homologous recombination during transfer of DNA into mouse L cells: role for DNA ends in the recombination process, *Mol. Cell. Biol.*, 4, 1020, 1984.

25. Waldman, A.S., Liskay, R.M., Differential effects of base-pair mismatch on intrachromosomal versus extrachromosomal recombination in mouse cells, *Proc. Natl. Acad. Sci. U.S.A.*, 84, 5340, 1987.

26. Waldman, A.S., Liskay, R.M., Dependence of intrachromosomal recombination in mammalian cells on uninterrupted homology, *Mol. Cell. Biol.*, 8, 5350, 1988.

27. Deng, C., Capecchi, M.R., Reexamination of gene targeting frequency as a function of the extent of homology between the targeting vector and the target locus, *Mol. Cell. Biol.*, 12, 3365, 1992.

28. Letsou, A., Liskay, R.M., Effect of the molecular nature of mutation on the efficiency of intrachromosomal gene conversion in mouse cells, *Genetics*, 117, 759, 1987.

29. Mansour, S.L., Thomas, K.R., Deng, C., Capecchi, M.R., Introduction of a lacZ reporter gene into the mouse int-2 locus by homologous recombination, *Proc. Natl. Acad. Sci. U.S.A.*, 87, 7688, 1990.

30. Shulman, M.J., Nissen, L., Collins, C., Homologous recombination in hybridoma cells: dependence on time and fragment length, *Mol. Cell. Biol.*, 10, 4466, 1990.

31. Hasty, P., Rivera-Pérez, J., Bradley, A., The length of homology required for gene targeting in embryonic stem cells, *Mol. Cell. Biol.*, 11, 5586, 1991.

32. Thomas, K.R., Deng, C., Capecchi, M.R., High-fidelity gene targeting in embryonic stem cells by using sequence replacement vectors, *Mol. Cell. Biol.*, 12, 2919, 1992.

33. Pennington, S.L., Wilson, J.H., Gene targeting in Chinese hamster ovary cells is conservative, *Proc. Natl. Acad. Sci. U.S.A.*, 88, 9498, 1991.

34. Yang, D., Waldman, A.S., An examination of the effects of double-strand breaks on extrachromosomal recombination in mammalian cells, *Genetics*, 132, 1081, 1992.

35. Valancius, V., Smithies, O., Double-strand gap repair in a mammalian gene targeting reaction, *Mol. Cell. Biol.*, 11, 4389, 1991.

36. Deng, C., Thomas, K.R., Capecchi, M.R., Location of crossovers during gene targeting with insertion and replacement vectors, *Mol. Cell. Biol.*, 13, 2134, 1993.

37. Mansour, S.L., Thomas, K.R., Capecchi, M.R., Disruption of the proto-oncogene int-2 in mouse embryo-derived stem cells: a general strategy for targeting mutations to non-selectable genes, *Nature*, 336, 348, 1988.

38. Hasty, P., Rivera-Pérez, J., Chang, C., Bradley, A., Target frequency and integration pattern for insertion and replacement vectors in embryonic stem cells, *Mol. Cell. Biol.*, 11, 4509, 1991.

39. Hasty, P., Rivera-Pérez, J., Bradley, A., The role and fate of DNA ends for homologous recombination in embryonic stem cells, *Mol. Cell. Biol.*, 12, 2464, 1992.

40. Adair, G.M., Nairn, R.S., Wilson, J.H., Seidman, M.M., Brotherman, K.A., MacKinnon, C., Scheerer, J.B., Targeted homologous recombination at the endogenous adenine phosphoribosyltransferase locus in Chinese hamster cells, *Proc. Natl. Acad. Sci. U.S.A.*, 86, 4574, 1989.

41. Zheng, H., Wilson, J.H., Gene targeting in normal and amplified cell lines, *Nature*, 344, 170, 1990.

42. Farzaneh, F., Panayotou, G.N., Bowler, L.D., Hardas, B.D., Broom, T., Walther, C., Shall, S., ADP-ribosylation is involved in the integration of foreign DNA into the mammalian cell genome, *Nucl. Acids Res.*, 16, 11319, 1988.

43. Waldman, B.C., Waldman, A.S., Illegitimate and homologous recombination in mammalian cells: differential sensitivity to an inhibitor of poly (ADP-ribosylation), *Nucl. Acids. Res.*, 18, 5981, 1990.

44. Waldman, A.S., Waldman, B.C., Stimulation of intrachromosomal homologous recombination in mammalian cells by an inhibitor of poly (ADP-ribosylation), *Nucl. Acids. Res.*, 19, 5943, 1991.

45. Voekel-Meman, K., Keil, L., Roeder, G.S., Recombination-stimulating sequences in yeast ribosomal DNA correspond to sequences regulating transcription by RNA polymerase I, *Cell*, 48, 1071, 1987.

46. Thomas, B.J., Rothstein, R., Elevated recombination rates in transcriptionally active DNA, *Cell*, 56, 619, 1989.

47. Nickoloff, J.A., Reynolds, R.J., Transcription stimulates homologous recombination in mammalian cells, *Mol. Cell. Biol.*, 10, 4837, 1990.

48. Wang, Q., Taylor, M.W., Correction of a deletion mutant by gene targeting with an adenovirus vector, *Mol. Cell. Biol.*, 13, 918, 1993.

49. Wang, Y., Maher, V.M., Liskay, R.M., McCormick, J.J., Carcinogens can induce homologous recombination between duplicated chromosomal sequences in mouse L cells, *Mol. Cell. Biol.*, 8, 196, 1988.

50. Mishina, Y., Ayusawa, D., Seno, T., Koyama, H., Thymidylate stress induces homologous recombination activity in mammalian cells, *Mutat. Res.*, 246, 215, 1991.

51. Radding, C.M., Homologous pairing and strand exchange in genetic recombination, *Annu. Rev. Genet.*, 16, 405, 1982.

52. Waldman, A., Unpublished.

Chapter 3

Gene Targeting in Trypanosomatids

Angela K. Cruz

Faculdade de Odontologia de Ribeirão Preto and
Faculdade de Medicina de Ribeirão Preto
Universidade de São Paulo
14049-900 Ribeirão Preto
São Paulo, Brazil

Contents

0-8493-8950-X/95/$0.00+$.50

I. Introduction

Trypanosomatids are eukaryotic organisms from the Order Kinetoplastidae, Trypanosomatidae Family. These unicellular eukaryotes are uniflagellate and invariably parasitic. Several kinetoplastids are of medical, veterinary, or agricultural importance because of their pathogenicity to man, domestic animals, and even crop plants. *Trypanosoma brucei* is the causative agent of sleeping sickness, a major disease in Africa. *T. cruzi* causes Chagas' disease, with 20 million people affected in Central and South America, and *Leishmania* are responsible for a disease that varies from a mild cutaneous to a visceral and lethal infection with a world-wide distribution.[1] Both genera are digenetic (heteroxeneous), alternating between two different animal hosts, a vertebrate and an invertebrate.

All kinetoplastids have a vesicular nucleus with a central nucleolus. They have a small genome of approximately 5×10^7 kb per haploid genome. Under electron microscopy it is possible to visualize perinuclear chromatin, but no chromosomal condensation cycle is apparent.[2] Thus, kariotype analysis was only made possible after development of pulse field electrophoresis that allowed the separation of chromosome-sized DNA molecules. According to that analysis, *T. brucei* chromosomes are separated in four classes depending on their size: the largest is over 10^6 to 10^7 bp and comprises 60% of the genome; a second class lies around 1000 to 2000 kb. There is a mid–range size (200 to 700 kb) and 100 minichromosomes of 50 to 150 kb range. The presence of these so–called minichromosomes is unique to the African trypanosome. *Leishmania* species have about 20 to 30 chromosomes; their sizes range between 200 and 2000 kb. It is a characteristic of the Order to have many small chromosomes rather than a few larger ones. Several laboratories have shown that there is a significant size polymorphism of certain chromosomes among species and also specie-specific variation in the size of some chromosomes.[3,4]

Trypanosomatid protozoa present some biochemical and molecular peculiarities, such as RNA editing of mitochondrial transcripts,[5] antigenic variation mediated by gene rearrangement, trans-splicing of a 39-base 5′-leader sequence onto protein-coding RNAs,[6] DNA amplification,[7-10] and bent DNA.[11] Many trypanosomatid genes are arranged in tandem repeats, and evidence suggests that expression involves multicistronic transcription and processing to yield mature mRNA.[12,13] These parasites contain unique organelles such as kinetoplast[14] and glycosome.[15] Thus, in addition to their medical importance, these parasites are also interesting model organisms for biochemical and molecular studies. Several studies have demonstrated the occurrence of genetic exchange between *T. brucei spp.* cotransmitted in tsetse flies in laboratory.[16-19] Sex is a comparatively rare event and is not an obligatory part of the trypanosome life cycle. Similar evidence has not been forthcoming for *T. cruzi*, but evidence for drug–resistant phenotypes suggests genetic recombination in *Crithidia fasciculata*.[20] Experimental crosses had not been successful in *Leishmania* species[21] although natural hybrids had been described.[22] As a consequence of the difficulties or impossibility to conduct conventional genetic crosses in trypanosomatids, classical genetics approaches could not be used, and progress on the molecular biology of the protozoan parasites has been hampered for many years.[23,24]

II. Transfer of genes into trypanosomatids

A. Historical background

It was not until 1989 that the powerful technique of DNA transfection was first used to conduct genetic studies in trypanosome and *Leishmania.*[25-30] The first successful DNA transfection experiments, using electroporation as a method for the introduction of the foreign DNA, had been reported in *Leptomonas* sp. (an insect trypanosomatide) and *Leishmania enrietti*. They were both described to be capable of transiently taking up and expressing exogenous DNA constructs containing genomic sequences fused to the bacterial gene CAT (chloramphenicol acetyltransferase).[25,28] The flanking parasite sequences used in these studies - mini-exon gene and putative polyadenylation signals from the α-tubulin[25] or intergenic sequences of the α-tubulin gene cluster[28] - were shown to contain the necessary signals for expression of the CAT gene, including proper processing of the primary transcript before translation by the host cell.

A couple of months later, the first stable transformants of *Leishmania* were obtained.[29,30] Laban et al.[29] replaced the CAT gene by a bacterial gene conferring resistance to neomycin (*neo*). After transfection of *L. enrietti* cells with the linearized plasmid (keeping complementary ends), they obtained transfectants that were resistant to neomycin and stable, if kept in the presence of the antibiotic. The plasmid was maintained as an extrachromosomal molecule. Kapler et al., based on their own previous work[7,8,31] on drug-induced gene amplification, developed a vector containing the so-called R region, which pops out of the chromosome (under methotrexate pressure) to generate a circular and stably maintained molecule bearing the dihydrofolate reductase-thymidylate synthase (DHFR-TS) and its flanking regions. Since the circle is associated with the overexpression of the DHFR-TS gene, it must contain all sequences responsible for control and appropriate gene expression. A bacterial plasmid backbone was added to the original circular R region, and DHFR-TS coding region was replaced with the neo cassette.

The best reported transfection efficiency[30] has been 10^{-4} transfected cells, which is comparable to those obtained for electroporation of DNA into cultured mammalian cells. The transfectants obtained had the exogenous DNA as an extrachromosomal particle. No *Leishmania*-stable transfectant bearing integration of exogenous DNA into the host cell genome was shown.

III. Targeted gene disruption

Gene targeting methodology has been used extensively in other organisms, ranging from yeast to mammalian cells.[32-34] It exploits the cellular machinery to mediate homologous recombination between exogenous and endogenous DNA. The use of such a technique would allow the modification of protozoan parasites' genome, creating specific alterations at a chosen locus. This reverse genetic technique would efficiently allow studies on gene function and expression in trypanosomatids.

The first reports on gene targeting in trypanosomatids were presented in 1990. It was known from work on other species[35] that linear targeting DNAs would favor integration of the targeting construct into the cell recipient genome, and that linear DNA with complementary ends should render transfectants with no genome integration.[29] In order to test the hypothesis, we transfected *L. major* cells with a linear DNA bearing non-complementary ends containing the neo gene joined to 2.2 kb of 3′- and 5.2 kb of 5′-flanking DNA corresponding to the endogenous DHFR-TS gene. Using low amounts of DNA, nearly 100% of the transfectants carried the planned homologous replacements.[36] The experiment clearly demonstrated that integrative transfection in *Leishmania major* resembles that of *Saccharomyces cerevisae* where homologous recombinants are the major integration product.

At the same time, the first report on homologous recombination by gene targeting was described in *T. brucei*.[37] They obtained stable *neo*^{r} transformants following the targeted insertion of the bacterial gene into the trypanosome tubulin cluster. The transfection vector used bears an in-frame fusion of the *T. brucei* tubulin gene and the *Escherichia coli neo*^{r} gene.

A. Methodological aspects

There are not many reports on gene targeting of protozoan parasites. Thus, it is possible to compile from each of these studies the methodology employed to obtain some general rules or possible approaches useful to other workers. Tables 1 and 2 summarize some of the methodological guidelines for targeting genes of trypanosomatids.

Cell lines to be transfected are grown to mid- or late-log phase and submitted to electroporation (electroporation conditions vary among different groups and genera or species to be transfected, Table 1).

Table 1. *Electroporation conditions*

Parasite	Cell stage	Cell conc. electrop. buffer	ZAP conditions	Screening of clones	Ref.
Leishmania enriettii	Mid-log phase	10^8 cells/ml HEPES based	3000V/cm 25 µF	Limiting dilution in microtiter plates	28,41
Leishmania major	Late log phase	10^8 cells/ml HEPES based	2250V/cm 500 µF	Isolated colonies from semisolid media	36,40, 51
Trypanosoma brucei	Log phase	$5{\times}10^7$ cells/ml phosphate based	250V, 500µF electrode gap?	Cloning medium + limited dilution	42
	Mid-log phase	10^8 cells/ml Zimmerman postfusion medium	4000V/cm 25 µF	No cloning	43
	Log phase (?)	(?)	4000V/cm 25 µF	No cloning	44
	Mid-log phase	$5{\times}10^7$ cells/ml Zimmerman postfusion medium	8000V/cm 25 µF	No cloning	37,56
	Mid-log phase	$5{\times}10^7$ cells/ml	300V, 250µF three pulses	No cloning	19
Trypanosoma cruzi	Mid-log phase	$8.6{\times}10^8$ cells/ml phosphate based	300V, 1500µF electrode gap?	Limiting dilution	45
	$5{\times}10^7$ cells/ml in LDNT medium	10^8 cells/ml phosphate buffer	60V/cm, 500 µF pulsed three times in close succession	neo^r cells from liquid cultures were spread into plates with G418. Colonies picked.	58

Features relevant to homologous gene replacement or integration at homologous sites in trypanosomatides are being gradually better understood with the increasing number of reports, gene targets, and approaches, such as using linear vs. circular molecules; targeting DNAs with homologous vs. nonhomologous ends or varying the length of homology between genomic sequence and targeting DNA; the amount of targeting DNA; the copy number of the target locus; and the length of the sequence to be replaced (e.g., how far the homologous sequences are from each other along the genome).

Table 2. *DNA form, amount and length of homology*

Parasite	Genomic target: replace (R) or insert (I)	*DNA* (μg)	Targeting DNA circular (C)/ linear (L)	Length of homology (kb)	R.E. digests	Events (insertion (I) vs. replacement (R))		Ref.
						R or I	R/I	
Leishmania								
enriettii	tubulin (I)	10	C	—	n.a.	no I	—	60
		10	L	0.2/0.8	single cut	I	—	
		5-25	L	0.9/1.0	N.C. ends	I	—	
		10	L	0.9/1.0	N.C. ends*	I	—	
	tubulin (R)	1-5	L	3.8/1.4	N.C. ends	R,I	22/17	41
Leismania	DHFR-TS**							
major	(R)	10	C	—	—	extra	—	36
		1-5	L	2.2/5.2	N.C. ends	R,I	13/1	
		10	L	2.2/5.5	N.C. ends	R,I	2/11	
		1-5	L	1.4/0.9	N.C. ends	R,I	5/1	
Trypanosoma	Calmodulin	20/ml						
brucei	(R)	(?)	L	6.0/6.0	N.C. ends*	R	27/0**	42
			C	—	n.a.	—		
	Ubiquitin (R)	high	C	—	n.a.	—	—	43
		2$	L	0.5/1.5	C. ends	R	—	
	Tubulin (I)	(?)	C	—	n.a.	—	n.a.	44
		(?)	L	0.3/0.3	C. ends	I	n.a.	
	Tubulin (I)	4.0	L	0.6/0.75	C. ends	I	n.a.	56
	Tubulin (I)	8.0	L	2.5/2.5	C. ends	R	—	37
		50.0	C	*ibid*	n.a.	I	—	
		8.0	L	0.5/0.5	C. ends	I	—	
		50.0	C	*ibid*	n.a.	—	—	
Trypanosoma	Calmodulin-							
cruzi	Ubiquitin (I)	200$	C	0.67/0.73	n.a.	I	n.a.	45
	TCR27 (R)	25	C	—	n.a.	—	—	58
			L	2.3/1.8	N.C. ends	R***		
			L	2.3/1.8	C. ends	R***		

* Targeting fragments carrying non complementary ends (two different restriction enzymes used), but not gel purified (at least not mentioned in the report).

$ Cells are incubated for 10 to 20 minutes with the exogenous DNA prior to electroporation. Carrying vector sequences (nonhomologous tail) on one side.

** Results from only seven clones are shown. Deletions of one to four copies of the calmodulin gene were observed.

*** Identical results from four "clones" are presented. It is not clearly indicated, however, from which electroporation experiment the clones come (linear DNA with or without complementary ends).

1. Length of homology

a. Gene replacement by homologous recombination — Gene targeting experiments using replacement vectors suggest that longer stretches of homology lead to higher frequencies of transfectants carrying a homologous gene replacement. We targeted the DHFR-TS gene in a clonal cell line of *L. major* with a linear targeting DNA-containing long stretches of homology between transfected fragment and chromosomal sequences (2.2 and 5.2 kb, respectively).[30,39] The DHFR-TS gene is a single-copy gene located in the 500 kb chromosome of *L. major* and is shown to be located in a diploid loci.[38] Our results indicate a surprisingly high frequency of homologous gene replacement that varies in the range of 20 to 100% with respect to the transfection efficiency of the same molecules in a circular form.[36] That frequency dropped to 5 to 10% when a targeting DNA carrying shorter DHFR-TS-flanking homologous sequences (0.9 and 1.4 kb of 5′- and 3′-flanking regions, respectively) was used[36,39,40] (Table 2).

No other data from trypanosomatids have been published on the influence of the extent of homology on the efficiency of homologous gene replacement. There are, however, two successful homologous gene replacements reported so far[41,42] (Table 2), where the authors present data from a considerable number of transfectants (more than 25 transfectants analyzed). The length of homology between targeting DNAs and genomic sequences were 1.4/3.8 kb or 6.0/6.0 kb on each side of the target locus. One of these studies shows that 56% of the transfectants obtained represented successful targeted deletions of an entire cluster of tubulin (20 or 40 kb long).[39,41]

b. Insertional targeting — It is clear from the compiled data (Table 2) that 300 base pairs of homology length between genomic and exogenous DNA sequences are enough to obtain integration at the specific genomic target. Transfectants bearing integration at the desired (homologous) genomic site were obtained in several studies with trypanosomes and *Leishmania* using such an extent of homology. It is not clear, however, how the extent of homology would affect the frequency of insertional events. All published reports show the analysis of just one or a few clones, and the length of homology is not changed. The primary interest resides on the generation of a given parasite mutant and not on the study of homologous recombination features in these protozoans. It is also not possible to make comparisons among different reports because of the variation of experimental conditions, such as electroporation parameters, amount of DNA, copy number of genomic targets, features of the DNA vector, or organism used.

2. Targeting DNA molecule

Trypanosomatids resemble *Saccharomyces cerevisiae* in that the preferred route of integration of a foreign DNA into the recipient genome is by homologous recombination (Table 2). All published studies confirm the complete lack of nonhomologous recombination and show that the observed events are exclusively homologous.

Regarding the use of circular molecules as transfecting DNA, it is a generalized observation among different groups that such circular molecules are either completely inefficient to generate transfectants[42-44] or they can persist in a few transfected cells as single or multiple arrays of the original vector. In two reports, however, integration of transfecting circular molecules into the genome of the recipient cells has been shown.[37,45] Such findings indicate that double-strand breaks may be required for efficient recombination in these parasites, similar to what is observed in yeast cells.[34]

In one case, where the integration of originally circular molecules in tandem arrays into the *T. cruzi* genome has been reported, a considerable amount of DNA (200 μg) was used and preincubated with the cells for 20 minutes prior to electroporation.[45] It can be speculated that such incubation would favor the action of endogenous nucleases that would create nicks and double breaks, generating a linear molecule and thus rendering the original targeting DNA an effective targeting agent.

As mentioned, circular molecules can eventually be integrated into these parasites' genomes by electroporation. However, the best approach to obtain a higher efficiency of either gene replacement or insertion into the genome of trypanosomes or *Leishmania* seems to be via electroporation with linear molecules.

The use of non-complementary ends in the targeting fragments to avoid recircularization might increase the frequency of integration events (either insertion or replacement) into the genome. None of the studies published so far have shown that the presence of complementary ends decreases the frequency of integration or replacement. Nevertheless, transfection of *L. enriettii* cells with a linearized targeting vector containing bacterial sequences on one of the extremities (one restriction enzyme, one single cut) generated only transfectants bearing extrachromosomal circular molecules.[29]

The relevance of using gel-purified vs. nonpurified targeting DNAs to the frequency of targeting events is not clear; authors do not always specify whether or not they gel purify the targeting fragments prior to transfection.

The influence of varying the amount of targeting DNA on the gene targeting efficiency has been shown in one report.[36] It is clear from the data presented that the higher frequencies (almost 100% of the events analyzed) of homologous gene replacement in *Leishmania* cells are obtained with low amounts of DNA (in the range of 1 to 5 μg). Such frequency drops to 45% of the events when higher amounts of DNA (10 μg) are transfected. In this case, the rest of the transfectants carry the targeting DNA integrated at the homologous locus.

By transfecting low amounts (1 to 5 μg) of a gel purified linear targeting DNA bearing non-complementary ends, Curotto de Lafaille and Wirth[41] have shown that replacement of an entire cluster of α-tubulin, representing a 20 to 40 kb long deletion from the genome, is possible. Clusters contain either 18 to 20 copies (large cluster) or 10 to 11 copies (small cluster) of the α-tubulin gene.

Because the cloning strategies used by the different groups for recovery and analysis of the transfectants can be diverse (Table 1), the recombination efficiency obtained is not always comparable.

Speculations on the possible effects of the copy number of the target sequence present in the genome on the observed frequency of gene targeting are not possible. Although it has been shown that the targeting frequency in mammalian cells is independent of the number of targets, it is definitively not true for yeast cells,[46] where gene targeting frequencies are proportional to the number of target sequences in the genome.

B. Gene targeting generates new data

1. Leishmania

A first round of transfection of a *L. major* cell line with a linear targeting DNA carrying the neo^r gene and DHFR-TS homologous sequences generates heterozygous clones (neo/+), by homologous replacement of one of the endogenous copies of the DHFR-TS gene by the neo^r gene. The other copy of the DHFR-TS gene can be targeted in a second round of transfection in a heterozygous clonal line (neo/+). In this case, the selectable marker neo^r on the targeting DNA is substituted by another selectable marker, the hygromycin B-resistance gene encoding hygromycin phosphotransferase (HYG). Targeting occurs with comparable efficiencies on both steps and independently of the order of the selectable markers.

We have used double gene targeting to create homozygous gene replacements. Although *Leishmania* are diploid at most loci,[36,38,47,48] they appear to be predominantly or exclusively asexual in nature and in the laboratory.[21,49,50] Thus, as we could not render an heterozygous parasite homozygous by sexual crossing, we used two independent selectable markers in successive rounds of gene targeting to replace both alleles of an endogenous gene. The DHFR-TS Leishmania obtained by double gene targeting represents a powerful tool for probing some of the unusual aspects of folate metabolism and methotrexate toxicity in this parasite. Phenotype analysis of these null mutants revealed that they are thymidine auxotroph.[40]

Engineering conditionally viable *Leishmania* by targeted deletion of essential metabolic genes may represent an interesting approach to generate safe vaccine lines to be used against leishmaniases. Since virulent *Leishmania major* may provide superior vaccination potential relative to attenuated lines we attempted to develop homozygous knockouts of DHFR-TS in virulent lines. To our surprise, we were unable to obtain the desired mutants, despite the ease with which such knockouts were obtained with attenuated lines. Instead, twice targeted virulent *Leishmania* cells resorted to extreme genetic measures to avoid complete loss of DHFR-TS. Most of the lines analyzed (9/11) underwent alterations in chromosomal number, either by aneuploidy or tetraploidy.[51]

These clones were shown to contain one or more normal DHFR-TS chromosomes and two modified chromosomes bearing neo and hyg replacements. We used all the experimental conditions known to recover null mutants for DHFR-TS of different organisms, including attenuated *Leishmania* clones.[40] Methodological problems excluded, the results indicated that DHFR-TS plays an additional unanticipated role in virulent lines, directly or in combination with other enzymes. Possible new roles for DHFR-TS virulent parasites is currently under investigation.

Many genes in the genome of these protozoan parasites are present in multiple copies organized in tandem arrays.[52-55] The deletion of a tubulin cluster demonstrated the feasibility of replacing these long sequences at once by homologous gene replacement[41] in *Leishmania enriettii*. After transfection of parasites with low amounts (1 to 5 μg) of a purified linear DNA containing the 5′- (5.8 kb) and 3′- (2.2 kb) flanking regions of the α–tubulin cluster, with non-complementary ends, seven out of eight transfections performed rendered chromosomal integration at the targeted locus. They have used two similar targeting vectors containing *neo* or *hyg* as selectable markers and, in consecutive transfections, tried to produce double deletion mutants. Two types of integration events were observed: in one type, *neo*^{r} and *hyg*^{r} replaced the complete tubulin cluster, and in the other type, the transfected gene integrated into the end of the original cluster. The null (*hyg/neo*) double-deleted mutants had lost the α–tubulin genes from the chromosome, but these genes remained as extrachromosomal elements. An interesting finding obtained from the functional analysis of these mutants is that, in those parasite cells, there is a post-transcriptional up-regulation in mRNA stability. Those up-regulated tubulin mRNA levels seemed to be sufficient for normal growth.

2. Trypanosome

a. T. brucei — A report[42] describes the targeting of calmodulin locus of *T. brucei*, with a calmodulin-*neo*^{r} fusion gene. The molecular karyotype analysis of 27 transformed cell lines showed that all of them presented integration into the genome by homologous recombination. No genomic integration was observed when circular targeting DNA molecules were used. In the cell lines analyzed, deletions of one to four copies of the calmodulin gene were observed.

Studies of different aspects of the biology of these trypanosomes are being carried out using gene targeting as a tool for understanding different phenomena observed in these parasites or in their genomic organization. An example is the study of the polyubiquitin locus of *T. brucei*.[43] The genomic organization of the locus has been studied by gene targeting. It was shown that *T. brucei* has a single polyubiquitin locus and that UbA and UbB are alleles that differ by the number of ubiquitin repeats they contain.

Gibson and Whittington[19] targeted the tubulin locus of trypanosome independently with two selectable markers to study the genetic exchange occurring in this parasite. Selection of hybrids by double drug resistance and analysis of the parasites from both fly midgut and salivary glands indicated the latter to be the site for mating. This is the first application of reverse genetics to the understanding of mating in trypanosome and will probably encourage new studies using the same approach.

Transcription in trypanosome is mediated by the same set of three RNA polymerases that have been identified in other eukaryotes. These RNA polymerases are distinguishable by their sensitivity to α-amanitin. Comparison of the sensitivity to α-amanitin of trypanosome protein-coding genes transcription revealed that some of them were transcribed by α-amanitin-resistant RNA polymerase.[26,27] Interestingly, they include the variant surface glycoprotein VSG-expression sites and the procyclic acidic repetitive protein PARP. This interesting feature of *T. brucei* is currently being explored by the gene targeting approach. Efficient gene transfer and targeted integration by homologous recombination led to the expression of the *neo*ʳ cassette, under the control of a PARP gene promoter. The selectable marker was targeted to an intergenic region in the α-tubulin-gene tandem array. The analysis of one transfectant suggests the presence of sixteen copies of *neo*ʳ in a tandem array. The PARP promoter controlled α-amanitin-resistant transcription of *neo*, whereas transcription of tubulin gene remained α-amanitin sensitive.[44]

The variant-specific surface glycoprotein (VSG) genes are responsible for the expression of a single type of protein that coats the *Trypanosoma brucei* surface. The organism has about 1000 different VSG genes, located at chromosomal internal positions or at telomeres. It is only from the telomeric loci on the larger chromosomes that a VSG gene may be expressed. The genes present at the silent genomic loci must be transposed to or recombined with the single active VSG gene expression site (ES) to be switched on. Trypanosome may activate a specific ES and inactivate a previously active one. The study of VSG genes expression and of the relevance of the specific chromosomal location of the ES promoter for the transcriptional activity has been approached with gene targeting technology.[56] A VSG-ES promoter element-driving *neo*ʳ gene was targeted into a polymerase II and chromosome-internal transcription unit: the tubulin gene array of procyclic trypanosome. To avoid readthrough transcription of the marker gene from tubulin promoter, the ES promoter was in inverse orientation relative to tubulin gene transcription. One single transformant was obtained with the *neo*ʳ gene in inverse orientation at the 5′ edge of the tubulin cluster. Low levels of expression of the selectable marker were observed. The experiment showed that expression of a VSG-ES promoter from an unusual location is possible. However, the transcription start point could not be located, nor could the sensitivity of transcription to α-amanitin be demostrated.

b. T cruzi — Although transient expression of *T. cruzi* was reported in 1991,[57] stable transformation was not successful in this trypanosome species until 1993. The first report[45] describes the stable transformation of *T. cruzi* resulting from the integration of a plasmid vector into targeted genomic sequences by homologous recombination. According to the authors, and unlike both *T. brucei* and *Leishmania*, this parasite does not require the use of linear transforming vectors. It was shown that a length of homology as short as 228 bp between ubiquitin coding sequence (the target locus) and plasmid sequences is enough for recombination to occur. They did not observe any readily identifiable phenotype corresponding to the disruption of the polyubiquitin gene.

The subject of another report using targeted gene replacement in *T. cruzi* was aimed at gaining insight into the function of a protein with a repetitive domain,[58] coded by the TCR27 gene. No recombinants were obtained with circular targeting DNAs. Only linear DNA, either excised insert or linear plasmid bearing non-homologous ends, generated homologous recombination events. Even though they could obtain the replacement of an entire TCR27 locus, they could not observe any phenotypic change caused by the genotypic alteration. Deletion of the second allele of the TCR27 gene was not possible.

Another report refers to the construction of a shuttle vector, which facilitates the expression of transfected genes in both *T. cruzi* and *Leishmania.*[59] Such a vector may prove to be a useful tool for testing genes and regulatory sequences from one trypanosomatid to the other and vice versa.

IV. Final remarks

Gene targeting methodology is generating important and also some unexpected data in the studies of trypanosomatids. The recently acquired ability to delete or truncate genes is bringing new insights. Understanding gene structure, genomic organization, and function of many different genes will be greatly accelerated by the new technology.

Leishmania genetic plasticity observed when deleting a DHFR-TS gene is an unexpected feature that shows, on one hand, that the existence of a novel role for this gene in virulent parasites compared to nonvirulent ones may be possible and, on the other hand, that the parasites' control of ploidy is highly unstable. The route chosen by *L. enriettii* cells targeted twice at the tubulin locus to bypass the lack of both chromosomal copies of the tubulin gene is to keep them as an extrachromosomal element.[41]

The studies of control elements and their specificities in *T. brucei* are complicated because chromosomes have their own control elements that may be interfering with the results obtained.[55] Existence of linear vectors, similar to the yeast artificial chromosomes (YAC) carrying telomeric sequences, would probably bring clearer responses to the addressed questions.

It may be also possible in the near future to assess the importance of homologous recombination for these parasites *in vivo*. It has been suggested that it would play an important role in the process of gene amplification.[60]

The use of this technology will foster rapid progress in the understanding of trypanosomatid biology itself, of parasite-host interactions, and of the diseases they cause.

Acknowledgments

The author thanks Professor Stephen M. Beverley and Luiz R. O. Tosi for helpful discussions during the preparation of this report.

References

1. WHO Expert Committee, The Leishmaniases, in *World Health Organization Technical Report Series*, Vol. 701, World Health Organization, Geneva, 1984, 3.

2. Vickerman, K., Phylum Zoomastigina, class Kinetoplastida, in *Handbook of Protoctista*, Margulis, L., Corliss, J.O., Melkonian, M., and Chapman, D.J., Eds., Jones and Bartlett Publishers, Boston, 1990, 215.

3. Lighthall, G.K. and Giannini, S.H., The chromosomes of *Leishmania*, *Parasitol. Today,* 8, 192, 1992.

4. Bastien, P., Blaineau, C. and Pages, M., *Leishmania*: sex, lies and karyotype, *Parasitol. Today*, 8, 174, 1992.

5. Simpson, L. and Shaw, J., RNA editing and the mitochondrial cryptogenes of kinetoplastid protozoa, *Cell*, 57, 355, 1989.

6. Borst, P., Discontinuous transcription and antigenic variation in trypanosomes, *Annu. Rev. Biochem.,* 55, 701, 1986.

7. Coderre, J.A., Beverley, S.M., Schimke, R.T. and Santi, D.V., Overproduction of a bifunctional thymidylate synthase-dihydrofolate reductase and DNA amplification in methotrexate-resistant Leishmania, *Proc. Natl. Acad. Sci. U.S.A.,* 80, 2132, 1983.

8. Beverley, S.M., Coderre, J.A., Santi, D.V. and Schimke, R.T., Unstable DNA amplifications in methotrexate-resistant *Leishmania* consist of extra-chromosomal circles which relocalize during stabilization, *Cell*, 38, 431, 1984.

9. Detke, S., Chaudhuri, G., Kink, J.A. and Chang, K.P., DNA amplification in tunicamycin-resistant *Leishmania* mexicana. Multiple copies of a single 63 kilobase supercoiled molecule and their expression, *J. Biol. Chem.*, 263, 3418, 1988.

10. Detke, S., Katakura, K. and Chang, K.P., DNA amplification in arsenite-resistant *Leishmania*, *Exp. Cell Res.*, 180, 161, 1989.

11. Marini, J.C., Levene, S.D., Crothers, D.M. and Englund, P.T., Bent helical structure in kinetoplast DNA, *Proc. Natl. Acad. Sci. U.S.A.,* 79, 7664, 1982.

12. Tschudi, C. and Ullu, E., Polygene transcripts are precursors to calmodulin mRNAs in trypanosomes. *EMBO J.*, 7, 455, 1988.

13. Muhich, M.J. and Boothroyd, J.C., Polycistronic transcripts in trypanosomes and their accumulation during heat shock: evidence for a precursor role in mRNA synthesis, *Mol. Cell. Biol.,* 8, 3837, 1988.

14. Simpson, L., The mitochondrial genome of kinetoplastid protozoa: genomic organization, transcription, replication and evolution, *Annu. Rev. Microbiol.*, 41, 363, 1987.

15. Opperdoes, F.R., Compartmentation of carbohydrate metabolism in trypanosomes, *Annu. Rev. Microbiol.*, 41, 127, 1987.

16. Jenni, L., Marti, S., Schweizer, J., Betschart, B., LePage, R.W.F., Wells, J.M., Tait, A., Paindavoine, P., Pays, E. and Steinert, M., Hybrid formation between African trypanosomes during cyclical transmisssion, *Nature*, 322, 173, 1986.

17. Sternberg, J., Tait, A., Haley, S., Wells, J.M., LePage, R.W.F., Schweizer, J. and Jenni, L., Gene exchange in African trypanosomes: characterization of a new hybrid genotype. *Mol. Biochem. Parasitol.,* 27, 191, 1988.

18. Gibson, W.C., Analysis of a genetic cross between *Trypanosoma brucei rhodesiense* and *T. b. brucei*. *Parasitol.*, 99, 391, 1989.

19. Gibson, W. and Whittington, H., Genetic exchange in *Trypanosoma brucei*: selection of hybrid trypanosomes by introduction of genes conferring drug resistance, *Mol. Biochem. Parasitol.,* in press.

20. Glassberg, J., Miyazaki, L. and Rifkin, M.R., Isolation and partial characterization of mutants of the trypanosomatid *Chrithidia fasciculata* and their use in detecting genetic recombination, *J. Protozool.*, 32, 118, 1985.

21. Panton, L.J., Tesh, R.B., Nadeau, K. and Beverley, S.M., A test for genetic exchange in mixed infections of *Leishmania major* in the sand fly *Phlebotomus papatasi*, *J. Protozool.*, 38, 224, 1991.

22. Evans, D.A., Kennedy, W.P., Elbihari, S., Chapman, C.J., Smith, V. and Peters, W., Hybrid formation within the genus *Leishmania*? *Parassitologia*, 29, 165, 1987.

23. Smith, D.F., Trypanosomatid transfection: stable introduction of DNA into protozoa, *Parasitol. Today*, 6, 245, 1990.

24. Borst, P. and Nussenzsweig, V., Molecular parasitology in Woods Hole, *Cell*, 71, 895, 1992.

25. Bellofatto, V. and Cross, G.A.M., Expression of a bacterial gene in a trypanosomatid protozoan, *Science*, 244, 1167, 1989.

26. Clayton, C.E., Fueri, J.P., Itzhaki, J.E., Bellofatto, V., Sherman, D.R., Wisdom, G.S., Vijayasarathy, S. and Mowatt, M.R., Transcription of the procyclic acidic repetitive protein genes of *Trypanosoma brucei*, *Mol. Cell. Biol.*, 10, 3036, 1990.

27. Rudenko, G., LeBlancq, S., Smith, J., Lee, M.G.S., Rattray, A. and van der Ploeg, L.H.T., Procyclic acidic repetitive protein (PARP) genes located in an unusually small alpha-amanitin resistant transcription unit: PARP promoter activity assayed by transient DNA transfection of *Trypanosoma brucei*, *Mol. Cell. Biol.*, 10, 3492, 1990.

28. Laban, A. and Wirth, D.F., Transfection of *Leishmania enriettii* and expression of chloramphenicol acetyltransferase gene, *Proc. Natl. Acad. Sci. U.S.A.,* 86, 9119, 1989.

29. Laban, A., Tobin, J.F., de Lafaille, M.A.C. and Wirth, D.F., Stable expression of the bacterial neo^r gene in *Leishmania enriettii*, *Nature*, 343, 572, 1990.

30. Kapler, G.M., Coburn, C.M. and Beverley, S.M., Stable transfection of the human parasite *Leishmania* delineates a 30 kb region sufficient for extra-chromosomal replication and expression, *Mol. Cell. Biol.,* 10, 1084, 1990.

31. Beverley, S.M., Ellenberger, T.E. and Cordingley, J.S., Primary structure of the gene encoding the bifunctional dihydrofolate reductase-thymidylate synthase of *Leishmania major*, *Proc. Natl. Acad. Sci. U.S.A.,* 83, 2584, 1986.

32. Capecchi, M., Altering the genome by homologous recombination, *Science*, 244, 1288, 1989.

33. Smithies, O., Gregg, R.G., Boggs, S.S., Koralewski, M.A. and Kucherlapati, R.S., Insertion of DNA sequences into the human chromosomal beta-globin locus by homologous recombination, *Nature*, 317, 230, 1985.

34. Orr-Weaver, T.L., Szostak, J.W. and Rothstein, R.J., Yeast transformation: a model system for the study of recombination, *Proc. Natl. Acad. Sci. U.S.A.,* 78, 635, 1981.

35. Orr-Weaver, T.L., Szostak, J.W. and Rothstein, R.J., Genetic applications of yeast transformation with linear and gapped plasmids, *Methods Enzymol.,* 101, 228, 1983.

36. Cruz, A. and Beverley, S.M., Gene replacement in parasitic protozoa, *Nature*, 348, 171, 1990.

37. Asbroek, T., Ouellette, M. and Borst, P., Targeted insertion of the neomycin phosphotransferase gene into the tubulin gene cluster of *Trypanosoma brucei, Nature*, 348, 174, 1990.

38. Iovannisci, D.M. and Beverley, S.M., Structural alterations of chromosome 2 in *Leishmania major* as evidence for diploidy, including spontaneous amplification of the mini-exon array, *Mol. Biochem. Parasitol.,* 34, 177, 1989.

39. LeBowitz, J.H., Coburn, C.M., McMahon-Pratt, D. and Beverley, S.M., Development of a stable *Leishmania* expression vector and application to the study of parasite surface antigen genes, *Proc. Natl. Acad. Sci. U.S.A.,* 87, 9736, 1990.

40. Cruz., A., Coburn, C.M. and Beverley, S.M., Double targeted gene replacement for creating null mutants, *Proc. Natl. Acad. Sci. U.S.A.,* 88, 7170, 1991.

41. Curotto de Lafaille, M. and Wirth, D.F., Creation of null/+ mutants of the Alfa-tubulin gene in *Leishmania enriettii* by gene cluster deletion, *J. Biol. Chem.*, 267, 23839, 1992.

42. Eid, J. and Sollner-Webb, B., Stable integrative transformation of *Trypanosome brucei* that occurs exclusively by homologous recombination, *Proc. Natl. Acad. Sci. U.S.A.,* 88, 2118, 1991.

43. Wong, S., Gottesdiener, K. and Campbell, D.A., Allelic polymorphism of the *Trypanosoma brucei* polyubiquitin gene conversion with a neomycin resistance marker, *Mol. Biochem. Parasitol.*, in press.

44. Lee, M.G. -S. and Van der Ploeg, L.H.T., Homologous recombination and stable transfection in the parasitic protozoan *Trypanosoma brucei, Science*, 250, 1583, 1990.

45. Capecchi, M., How efficient can you get?, *Nature*, 348, 109, 1990.

46. Hariharan, S., Ajioka, J. and Swindle, J., Stable transformation of *Trypanosoma cruzi*: inactivation of the PUB12.5 plyubiquitin gene by targeted gene disruption, *Mol. Biochem. Parasitol.,* 57, 15, 1993.

47. Leon, W., Fouts, D.L. and Manning, J.M, Sequence arrangement of the 16S and 26S rRNA genes in the pathogenic haemoflagellate *Leishmania donovani*, *Nucl. Acids Res.*, 5, 491, 1978.

48. Iovannisci, D.M., Goebel, D., Allen, K., Kaur, K. and Ullman, B., Genetic analysis of adenine metabolism in *Leishmania donovani* promastigotes, *J. Biol. Chem.,* 259, 14617, 1984.

49. Tait, A., Sexual processes in the kinetoplastida, *Parasitology*, 86, 29, 1983.

50. Tibayrenc, M., Kjellberg, F. and Ayala, F.J., A clonal theory of parasitic protozoa: the population structures of *Entamoeba*, *Giardia*, *Leishmania*, *Naegleria*, *Plasmodium*, *Trichomonas*, and *Trypanosoma* and their medical and taxonomic consequences, *Proc. Natl. Acad. Sci. U.S.A.,* 87, 2414, 1990.

51. Cruz, A.K, Titus, R., Beverley, S.M., Plasticity in chromosome number and testing of essential genes in *Leishmania* by targeting, *Proc. Natl. Acad. Sci. U.S.A.,* 90, 1599, 1993.

52. Seebeck, T., Wittaker, P.A., Imboden, M.A., Hardman, N. and Braun, R., Tubulin genes of *Trypanosoma brucei*: a tightly clustered family of alternating genes, *Proc. Natl. Acad. Sci. U.S.A.,* 80, 4634, 1983.

53. Thomashow, L.S., Milhausen, M., Rutter, W.J. and Agabian, N., Tubulin genes are tandemly linked and clustered in the genome of *Trypanosoma brucei, Cell,* 32, 35, 1983.

54. Landfear, S.M. and Wirth, D.F., Tandem arrangement of tubulin genes in the protozoan parasite *Leishmania enrietti, Mol. Cell. Biol.*, 3, 1070, 1983.

55. Huang, P.L., Roberts, B.E., McMahon-Pratt, D., David, J.R. and Miller, J.S., Structure and arrangement of the beta-tubulin genes of *Leishmania tropica, Mol. Cell. Biol.*, 4, 1372, 1984.

56. Zomerdijk, J.C.B.M., Kieft, R. and Borst, P., Insertion of the promoter for a variant surface glycoprotein gene expression site in an RNA polymerase II transcription unit of procyclic *Trypanosome brucei, Mol. Biochem. Parasitol.*, 57, 295, 1993.

57. Lu, H.Y. and Buck, G.A., Expression of an exogenous gene in *Trypanosoma cruzi* epimastigotes, *Mol. Biochem. Parasitol.,* 44, 109, 1991.

58. Otsu, K., Donelson, J.E. and Kirchhoff, L.V., Interruption of a *Trypanosoma cruzi* gene encoding a protein containing 14-amino acid repeats by targeted gene insertion of the neomycin phosphotranferase gene, *Mol. Biochem. Parasitol.,* 57, 317, 1993.

59. Kelly, J.M., Ward, H.M., Miles, M.A. and Kendall, G., A shuttle vector which facilitates the expression of transfected genes in *Trypanosoma cruzi* and *Leishmania, Nucl. Acids Res.*, 20, 3963, 1992.

60. Tobin, J.F. and Wirth, D.F., A sequence insertion targeting vector for *Leishmania enriettii, J. Biol. Chem.*, 267, 4752, 1992.

Chapter 4

Gene Targeting in Plants

Remko Offringa and Paul Hooykaas

Institute of Molecular Plant Sciences,
Leiden University, Clusius Laboratory,
Wassenaarseweg 64,
2333 AL Leiden, The Netherlands

Contents

0-8493-8950-X/95/$0.00+$.50

I. Introduction

The development of methods for the introduction of exogenous DNA into plant cells and the use of transgenic plants have been major steps forward in plant molecular biology. It has contributed to the study of the regulation of plant gene expression by using promoter-marker gene fusions and, moreover, it has allowed the study of gene function *in vivo* via overexpression of genes in either sense or antisense orientation. The knowledge of developmental and morphogenetic processes in plants is rapidly increasing, and metabolic routes are being unravelled. This has been aided by the identification of genes responsible for these processes in plants. In plant bio-technology, transgenesis has allowed the engineering of new genetic traits in crop plants, such as herbicide tolerance,[1] prolonged shelf-life of fruit,[2] altered oil composition in seeds,[3] and increased resistance to insects[4] or diseases caused by bacterial,[5] viral,[6] or fungal[7] pathogens.

Exogenous DNA that is introduced into plant cells is efficiently integrated as part of the plant genome. As in mammalian cells, the site of insertion cannot be predicted since integration occurs via nonhomologous (illegitimate) recombination.[8,9,10] This may cause problems such as unreliable expression of a transgene or unwanted mutagenesis by the disruption of a gene at the site of insertion. Gene targeting, i.e., the site-directed insertion of exogenous DNA into the genome via homologous recombination, should overcome these problems. More importantly, as gene targeting has been justified as an essential tool for reverse genetics in mammalian systems, it will most likely also prove the same for plant systems.

In this chapter, the different methods for introduction of exogenous DNA in plant cells will be briefly discussed, and the different studies on homologous recombination in plants will be presented. Subsequently, we will discuss the first experimental results concerning gene targeting in plants and conclude with future perspectives for the site-directed mutagenesis of the plant genome.

II. DNA transfer to plants

Several methods have been developed for the introduction of exogenous DNA into plant cells. With most of the earlier methods, purified DNA was introduced into plant cells that lack their cell wall (protoplasts) via either polyethylene glycol (PEG) treatment,[11] electroporation,[12] fusion with liposomes,[13] or micro-injection.[14] For some plant species, the regeneration of transgenic lines from single plant cells is relatively simple, and progeny analysis is easy since large numbers of seeds can be obtained from one plant. A recently developed technique is the introduction of DNA via particle bombardment into whole plant tissues,[15] thus avoiding problems usually occurring with the regeneration of transgenic plants from single cells. Most recently it also appeared to be possible to transform whole plant tissues via electroporation of DNA.[16]

In addition to these methods, the bacteria *Agrobacterium tumefaciens* and *Agrobacterium rhizogenes* are used as efficient vectors for transformation of many dicotyledonous plant species.[17] Both *Agrobacterium* species are capable of transferring a defined piece of DNA, the transferred DNA (T-DNA), to the plant cell. Not only protoplasts,[18] but also cells that are part of whole plant tissues such as segments from leaves, tubers, or roots can be transformed via cocultivation with *Agrobacterium.*[19–21] For a long time, only dicotyledonous plants were believed to be recipients for T-DNA transfer. However, since the initial observations in *Chlorophytum* and *Narcissus*[22] *Agrobacterium*-mediated DNA transfer has been achieved in more and more monocot species.[23–25]

The mechanism of T-DNA transfer has been the subject of detailed studies, which have been extensively reviewed.[26,27] Two genetic elements that are present on a large plasmid in *Agrobacterium,* the T-region and the virulence (vir) region, are essential for T-DNA transfer. The T-region is the DNA segment that is transferred to the plant cell. The vir region encodes proteins that form the core of the transfer machinery. The T-region is bordered by imperfect 24 bp direct repeats, which are the only cis-acting sequences that are essential in the transfer process. The rest of the T-region can be replaced by DNA sequences of interest. In order to simplify manipulation via molecular cloning techniques, the T-region is used in trans with respect to the vir region on a plasmid (binary vector) that can be shuttled between *Agrobacterium* and *E. coli.* The border repeats are recognized by VIRDl and VIRD2, two virulence proteins encoded by the virD operon. The cooperative action of these proteins results in bottom strand nicks between bases 3 and 4 of the repeats. Subsequently, single-stranded (ss) copies (T-strands) of the T-region bottom strand are generated via replacement synthesis, utilizing the border nicks as initiation and termination sites (Figure 1). The VIRD2 protein remains covalently linked to the 5′ terminus of the T-strand and VIRE2 proteins bind to the T-strand, thus forming the T-complex. The T-complex is exported to the plant cell via a pore in the bacterial membrane, which consists of proteins encoded by the virB operon. Nuclear localization signals present in both VIRD2 and VIRE2 probably direct transport of the T-complex to the nucleus of the plant cell.[28–31] Most likely, the role of the protein coat of the T-DNA transfer intermediate is to protect the T-strand from nuclease activity during transport. In contrast with transformation methods that use purified DNA, *Agrobacterium*-mediated transformation typically results in low copy T-DNA inserts, which are perfect (non-rearranged) copies of the T-region.[32–34]

Using particle bombardment or PEG treatment, it has also been possible to introduce exogenous DNA into the chloroplast of plant cells.[35] The introduced DNA is predominantly inserted via homologous recombination as in bacteria[36] and lower eukaryotes.[37,38] Chloroplast transformation via *Agrobacterium* has been reported.[39,40]

III. Homologous recombination in plants

The molecular basis of DNA-repair, recombination, and more specifically, homologous recombination between DNA molecules have been sources of intensive study in bacteria and lower eukaryotes (see Chapters 1, 2, and 3). In the last decade, a great deal of information has also been gathered concerning these processes in mammalian cells (see Chapters 1, 2, 5, and 6). In plants, however, more detailed molecular studies were commenced only recently.

The first evidence that homologous recombination occurs in mitotically active plant cells came from studies with mutant plant viruses. Plants co-infected with mutant cauliflower mosaic virus (CaMV) genomes having non-overlapping mutations accumulated virus particles, which were produced by recombination.[41,42] In 1987, Wirtz and co-workers[43] were the first to provide molecular evidence for homologous recombination between co-introduced non-viral plasmids in plant cells. Soon after that, Paszkowski and co-workers[44] reported the first evidence of gene targeting in plants. In this study, targeted recombination between a defective Kanamycin resistance (Kmr) gene residing at a chromosomal locus and an introduced repair construct was detected in approximately 1 out of 10^4 transformants.

Different studies have been initiated to gain a better understanding of the process of homologous recombination in plants. A summary of the studies of extrachromosomal recombination (ECR, i.e., between co-introduced DNA molecules) and intrachromosomal recombination (ICR, i.e., between closely linked chromosomal repeats) will be presented below, and results will be

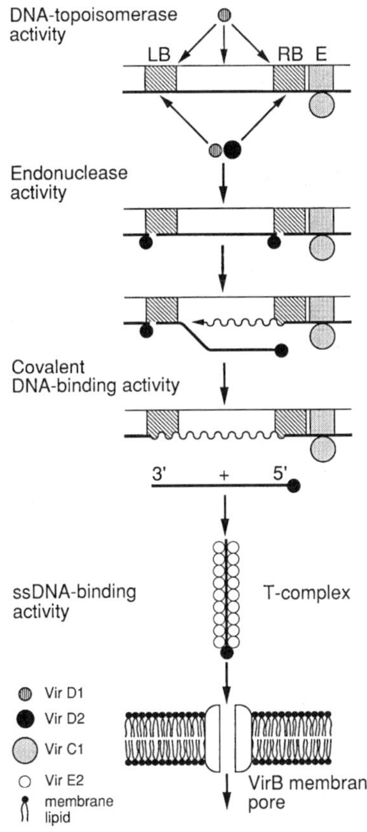

Figure 1. *Model for formation of the DNA intermediate that is transferred from* Agrobacterium *to the plant cell. Cooperative action of Vir D1 (has topoisomerase activity) and Vir D2 results in bottom strand nicks at the right and left border repeat (**RB** and **LB**, respectively). Vir D2 remains covalently linked to the 5′ terminus of the nicked DNA and single-stranded copies of the T-region (T-strands) are produced via replacement synthesis. Vir E2 binds to the T-strand and the resulting DNA::protein complex (T-complex) is transported via the Vir B membrane pore and a pilus-like structure to the plant cell. The overdrive or enhancer sequence (**E**) located next to the RB stimulates border-mediated T-DNA transfer probably via binding of the Vir C1 protein.*

compared to results in mammalian systems. Subsequently, the first experiments on gene targeting in plants will be discussed, followed by suggestions to make the method generally applicable to plant systems.

A. Extrachromosomal homologous recombination (ECR)

Homologous recombination can be studied in higher eukaryotes by co-introducing DNA molecules that have a region of homology. The recombination process is believed to occur extrachromosomally, i.e., before the DNA molecules become part of the chromosome structure, since it occurs with a very high frequency compared to recombination between two repeats at one chromosome (ICR) or between a chromosomal and an introduced DNA homologue (gene targeting).[45] The high efficiency of the process has made ECR an attractive experimental system in mammalian cells as well as in plant cells.

ECR has been studied in protoplasts of *Nicotiana tabacum*[43,46–48] and *Petunia hybrida*[49] by co-introduction of two plasmids, each containing a defective derivative of an antibiotic resistance gene with non-overlapping mutations and selection for resistant calli having the restored selection marker inserted. Typically, the product of recombination was detected in 1 to 30% of the transformants. In more recent approaches, homologous recombination between such constructs was monitored by measuring transient expression of a restored marker gene in protoplasts of *Nicotiana tabacum,*[47] *Nicotiana plumbagenifolia,*[50–52] and the monocot *Zea mays.*[53] The occurence of homologous recombination was confirmed at the DNA level using direct PCR amplification.[50,53]

The efficiency of recombination is positively correlated with the length of the overlapping homology, referred to as the effective homologous region (EHR).[46,50,53] The homology required for efficient homologous recombination in plants is similar to that required in mammalian systems[50] (see Chapters 1 and 2). The topological state of the plasmids [i.e., supercoiled vs. linearized by a double strand break (DSB)] influences the recombination frequency. The position of the DSB also appears to be important. In general, the highest recombination frequencies are obtained when the recombination substrates are linearized. This stimulating effect is only observed when the EHR is kept at maximal size. Most products of ECR are formed within 30 minutes after DNA transfer,[52] which is similar to the time reported in mammalian cells.[54]

Until now, only a few groups have made a serious attempt to study the mechanism of recombination in plant cells.[47,48,51] In mammalian cells, the single strand annealing (SSA) model has been used to explain extrachromosomal recombination instead of the double strand break repair (DSBR) model (see Chapters 1 and 2). One of the basic differences between the two models is that DSBR requires a DSB in only one of the recombination partners, whereas SSA requires a DSB in both DNA molecules. In the latter case, DSBs have been proposed to be entry sites for single strand (ss) exonucleases that produce complementary single DNA strands.[55–57] Alternatively, ssDNA ends may be generated by a helicase that unwinds dsDNA.[58] After its formation, complementary ssDNAs of the different plasmids have the opportunity to anneal and form a recombination product. This process is referred to as being non-conservative, since the opposite ends of the recombining plasmids that are not involved in the exchange process suffer degradation. The model implies that limited non-homology at the ends of the recombination substrates does not have a severe effect on the efficiency of recombination.[55,58] This is different from DSBR, which is a conservative reaction (i.e., all ends are retained) and requires a DSB in the region of homology for efficient recombination.[59]

In plants, as in mammalian systems, ECR seems to occur predominantly via the SSA model,[46,47,48,51] although de Groot and co-workers[48] demonstrated that ECR in plant cells can also occur via the DSBR pathway. Linearization outside the region of homology, which results in a small non-homologous region at the ends of the DNA molecule, does not have a dramatic effect on the efficiency of recombination in tobacco.[46,51] In corn, however, the ECR reaction is severely impaired by small regions of non-homology at ends of the EHR.[53] This could indicate a difference of exonuclease/helicase contents in monocot compared to dicot cells. The role of single-stranded DNA in intermolecular recombination has been implicated by experiments performed by two groups.[47,48] The experiments of Bilang et al.[47] showed that SSA is an important step in the recombination reaction and that the exposure of a homologous ss stretch in dsDNA is likely to be a rate-limiting step in recombination. De Groot and co-workers[48] demonstrated a slight but significant enhancement of the recombination efficiency when one of the dsDNA substrates was replaced by ssDNA. In this case, ECR occurred equally efficiently via the SSA or the DSBR pathway.

B. Intrachromosomal homologous recombination (ICR)

An alternative way to study homologous recombination in plant cells is to use artificial, closely linked chromosomal repeats. This is referred to as intrachromosomal homologous recombination (ICR). The study of ICR may help to explain processes like genetic drift, genomic rearrangements, and instability of transgenes. The first experiments were performed in *Nicotiana tabacum* and *Arabidopsis thaliana*. In these studies, the repeats consisted of complementary defective derivatives of the kanamycin resistance (Kmr) gene (neomycin phosphotransferase [nptII] coding region between plant transcription regulator sequences) having non-overlapping mutations. These derivatives were present either in direct[60–62] or inverted[63] orientation on a DNA construct that was stably integrated in the nuclear plant DNA. Somatic ICR was detected in cells of selected plant lines by scoring kanamycin-resistant calli regenerating from protoplasts of these parental lines. To assess the efficiency of ICR after meiosis, parental transgenic lines were self-pollinated and seedlings were tested for kanamycin resistance.

An alternative system to detect ICR was used in *Brassica napus* (rapeseed). Here a partial dimer of the CaMV genome was inserted into the nuclear DNA such that an infectious CaMV circle could only be produced via ICR at the repeated sequence.[64,65] Excision of CaMV could be visually scored since viral infection produces chlorotic plaques and vein clearing on leaves. The 1033-bp homologous regions used in these studies were derived from two different CaMV strains and had 34 base pair differences, which allowed the detection of the site of recombination. Viral DNAs were isolated from progeny of the original plant line and the homologous region of recombination was analyzed.

In general, the frequency of ICR in somatic cells varies between 10^{-4} and 10^{-6} per cell clone,[60,61,63] which is similar to the frequencies of ICR observed in mammalian cells[45] and much lower than the frequencies of ECR. The average length of the EHR in the test constructs is 400 bp, but 53 bp appears to be enough to get detectable ICR in somatic cells.[60] The efficiency of ICR differs among plant lines and seems to be dependent on the position and/or the structure of the locus.[61,63,65] Per locus, the frequency of ICR seems to be independent of the presence of non-allelic loci.[66] However, when a plant line contains several artificial repeats in tandem at one genomic location, multiple ICR events occur at this position at a relatively high frequency.[60,61] Thus, the occurence of an ICR event seems to stimulate recombination between repeats at closely linked positions. The rate of ICR in progeny homozygous for a test construct containing a direct repeat is twofold higher than in the hemizygous parent lines.[62,65,66] This is to be expected, since homozygous lines contain two copies of the test construct and, thus, have two positions where ICR may occur. Tovar and Lichtenstein[63] reported that ICR is more than twofold higher in plants homozygous for a test construct having inverted repeats instead of direct repeats. The results suggest that in these plants, homologous recombination occurs between the repeats in test construct, as well as between test constructs at allelic postions via unequal pairing of the sister chromatids.

A general problem when studying ICR in seedlings is that resistance can arise from either somatic (either in the parent plant or in the early embryo) or germinal ICR events. Among a total of 3×10^4 seedlings, Peterhans and co-workers[61] observed seedlings with resistant patches of tissue (Figure 2), but fully green seedlings indicative of germinal events were not obtained. The results provide direct evidence for *in planta* ICR in somatic tissue and indicate that the frequency of ICR in germinal tobacco cells is below 3.3×10^{-5}. In similar experiments, Tovar and Lichtenstein[63] found that 2 of the 3×10^5 seedlings tested were fully green kanamycin resistant, indicating an ICR frequency of 6.7×10^{-6} in germinal cells. Similar experiments with transgenic *Arabidopsis thaliana* plants containing one insert of an artificial direct repeat provided fully green-resistant seedlings, as well as seedlings with green-resistant sectors.[61] These results suggested that germinal events may occur at a frequency of about 10^{-6}.

ICR occurs with high fidelity,[60] and in the experiments of Gal and co-workers,[64] results suggest that stretches of heteroduplex DNA are resolved by mismatch repair. Nonreciprocal recombination (i.e., gene conversion) seems to be the preferred mechanism of ICR.[61,63] Only 2 of 28 recombination events observed by Tovar and Lichtenstein[63] between inverted repeats were reciprocal and, interestingly, in lines homozygous for the test construct, a small frame shift mutation in the Kmr gene (npt^-) was corrected more frequently than a larger deletion (npt^D). Moreover, co-conversion events (i.e., repair of both npt^- and npt^D) were also detected in both hemizygous and homozygous lines, and can be accounted for by mismatch repair of an ICR intermediate or via DSBR during unequal pairing of the sister chromatids.

Several groups have tested treatments that may induce ICR in plant cells. Peterhans and co-workers[60] have observed that protoplast culture, by itself, does not induce ICR. However, ICR can be induced in somatic cells by low dose ionizing irradiation,[62,64] treatment of plant cells with mitomycin-C (MC), or heat shock.[62] The MC treatment seems to be the most promising, since it

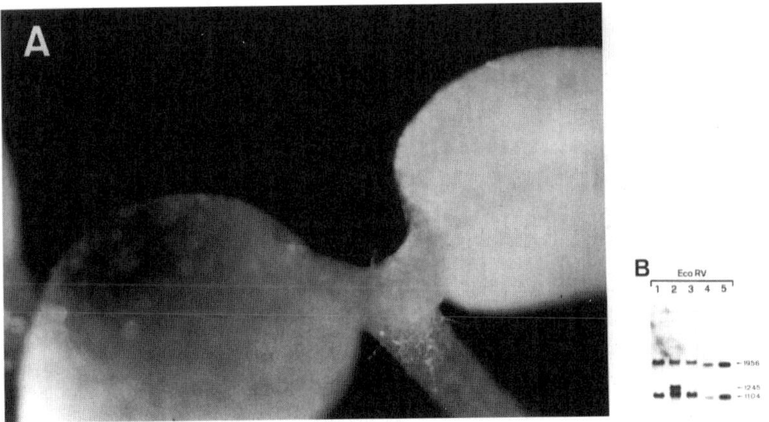

*Figure 2. Intrachromosomal recombination in plants. (**A**) Seeds obtained after self pollination of a transgenic tobacco line that contains a test construct for ICR (consisting of a direct repeat of two complementary defective Kmr genes) were germinated in the presence of kanamycin sulfate. A green area of tissue on the bleached cotelydons of a seedling indicated the formation of a functional Kmr gene via ICR. (**B**) Southern blot analysis shows a 1245 bp fragment indicative for the product of ICR in a cell population derived from the green area and grown with (lane 2) or without (lane 3) kanamycin selection. Lane 1, parental plant line; Lane 4 and 5, 10 and 30 pg of a plasmid containing the test construct. From Peterhans, A., Schlüpmann, H., Basse, C., and Paszkowski, J., EMBO J., 9, 3437, 1990. By permission of Oxford University Press.*

gives the greatest increase (ninefold) in ICR without affecting cell viability.[62] This is in contrast to mammalian cells, which are already sensitive to MC at relatively low concentrations and show only limited stimulation of ICR by this compound.[61] This could indicate a lower threshold for induction of recombinational repair by MC in plant cells than in mammalian cells.

C. ECR vs. ICR as model for gene targeting in plants

A major concern is whether findings from ECR experiments are valuable to elucidate the mechanism of gene targeting in plants. If we consider results in mammalian systems, it is doubtful whether ECR reflects recombination during gene targeting where one of the sequences is already part of the chromatin structure (see Chapters 1 and 2). ECR occurs with relatively high efficiency, is not drastically reduced by mismatches,[68] and is thought to occur predominantly via the nonconservative SSA pathway.[47,48,51,55,56,58] In contrast, ICR involves DNA homologies that are part of the chromatin structure, is found to occur at much lower frequencies, is strongly affected by base-pair mismatches, and follows the conservative DSBR pathway.[45,60] ICR may, therefore, be a better source of information on the mechanism of gene targeting.[45]

IV. Gene targeting in plants

A. Gene targeting using direct DNA transfer

Until recently, little experimental data concerning gene targeting in plants have been published. The first demonstration of targeted recombination between introduced plasmid DNA and a chromosomal DNA sequence in mitotically active plant cells was reported by Paszkowski and co-workers.[44] Protoplasts of five different transgenic lines of *Nicotiana tabacum*, each having two or more copies of a mutant Kmr gene inserted at one location in the nuclear genome, were transformed via PEG treatment with a repair construct containing a defective Kmr gene with a non-overlapping mutation. The EHR between the chromosomally located and the repair construct was 352 bp. Seven kanamycin-resistant calli were obtained from one plant line, and one Kmr callus was obtained from another plant line. Further analysis showed that the formation of the intact Kmr gene was a result of homologous recombination between repair construct and target. Targeted recombination had occurred in approximately one out of 10^4 transformed cells, which is much less frequent than recombination between co-introduced DNA copies (ECR).

More recently, gene targeting was also achieved in cells of *Arabidopsis thaliana*.[69] A chimeric hygromycin-resistance (Hmr) gene, consisting of the coding region of the bacterial hygromycin phosphotransferase gene (hpt) between plant transcription regulator sequences, was made defective by a 19-bp deletion in the coding region. This construct was first introduced as artificial target using *Agrobacterium*-mediated transformation. Eleven transgenic lines were selected, harboring between one and seven copies of the target construct, and protoplasts of these lines were transformed with a targeting construct containing the intact coding region of the hpt gene, providing an EHR of 1050 bp. The DNA was introduced via PEG treatment either as isolated fragment, as linearized plasmid, as linearized plasmid together with carrier DNA, or as single-strand circular DNA.[70] In 4 out of 150 hygromycin-resistant calli that were obtained, the presence of an intact gene was detected by PCR analysis. The occurrence of gene targeting could be demonstrated in two recombinants, each obtained from a different target line. One of the calli was lost due to infection. The fourth recombinant clone derived from the same target line appeared to be a false positive after detailed analysis of the target locus. In this clone, an unchanged targeting construct appeared to have integrated close to the target locus. Most likely, a correct PCR fragment indicative for the presence of the corrected hpt gene was obtained due to template switching during the PCR.[71] This underlines the possible occurrence of recombination during PCR, thereby creating artefacts in the detection of targeted recombination. Gene targeting in these experiments was detected in 1 out of 3×10^4 transformants, which is comparable to the results obtained by Paszkowski and co-workers.[44]

B. *Agrobacterium* — a suitable vector for gene targeting in plants

Initially the *Agrobacterium* vector system seemed not very suitable for gene targeting in plants, since the binding of VIRD2 and VIRE2 proteins to the T-DNA (see above) could possibly hamper its interaction with homologous DNA sequences in the plant genome. To test this, we co-cultivated tobacco protoplasts with two *Agrobacterium* strains, each harboring a different but homologous T-DNA construct. ECR between the two T-DNA constructs was detected in 1 to 4% of the co-transformed tobacco cells, demonstrating that T-DNA is a suitable substrate for homologous recombination in plant cells.[72] Moreover, two years after the first report on gene targeting in plants,[44] Lee and co-workers[73] and our group[72] independently reported on the use of *Agrobacterium* as a vector for gene targeting in plants.

Lee et al.[73] were the first to use an endogenous plant gene as target. This target gene encodes acetolactate synthase (ALS), which performs the first common step in the biosynthesis of branched-chain amino acids in plants. ALS is the site of action for the sulfonylurea herbicides,[74] and a TrpS[73] to Leu amino acid change in the protein confers resistance to the herbicide chlorsulfuron.[75] *Nicotiana tabacum* contains two genetically unlinked and most likely homologous ALS loci, SurA and SurB.[76] A 1.9-kbp fragment containing the 3′ part of the SurB gene with the chlorsulfuron resistance mutation was used as targeting construct. An Apa I restriction site introduced 25 bp from the chlorsulfuron resistance mutation enabled the distinction between targeted recombinants and spontaneous mutations. Seven million tobacco protoplasts were co-cultivated with an *Agrobacterium* strain carrying the targeting vector, and after chlorsulfuron selection, seven resistant calli were obtained. Four of the colonies appeared to be spontaneous mutants, but in the other three, the product of homologous recombination between targeting vector and the identical SurB locus could be detected, the relative frequency of targeted recombination being 8.4×10^{-5}.

The experimental design used in our group (see Figure 3) was comparable to that of Paszkowski et al.[44] and Halfter et al.[70] A T-DNA construct containing an Hmr gene and a 3′ deleted Kmr gene was used to create an artificial target. Protoplasts of plant line T, which is hemizygous for a target locus harboring two of these T-DNAs in an inverted orientation, were used for transformation with a T-DNA repair construct consisting of a Hmr gene and a defective Kmr gene having its 5′ part deleted. Kmr was obtained at a frequency of 1:500 transformed cells, but when 213 kanamycin-resistant clones were analyzed by PCR, the product of homologous recombination could only be detected in three of them. Southern blotting and sequence analysis revealed that the defective Kmr at the target locus had been accurately restored in one of the recombinant clones, clearly demonstrating the occurence of gene targeting. Stable transmission of the targeted locus could be demonstrated after crossing the recombinant line with nontransgenic tobacco (Figure 4).

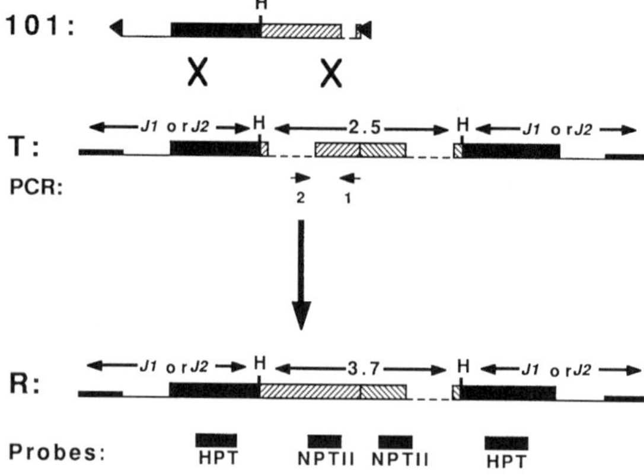

Figure 3. *Restoration of a defective gene copy via* Agrobacterium-*mediated gene targeting. The target line (**T**) contains two T-DNA inserts that are integrated in an inverted repeat structure. Each T-DNA insert carries a hygromycin resistance gene (black bar) and a defective kanamycin resistance gene (striped bar) from which the 3′ part has been deleted (dotted line). Protoplasts of the target line were co-cultivated with the* Agrobacterium *strain containing the binary vector with repair T-DNA.*[101] *The repair construct carries a defective kanamycin resistance gene from which the 5′ part has been deleted. The black triangles indicate the border repeats. Kanamycin-resistant calli were regenerated and analyzed by PCR (primers 1 and 2) for the presence of a restored gene. Plants were regenerated from the positive calli and Southern blot analysis was performed to test for the gene targeting event. A shift of the NPTII hybridizing HindIII fragment from 2.5 to 3.7 kbp is indicative for the targeting event. The HPT hybridizing junction fragments J1 and J2 are left unchanged. A recombinant line (**R**) was identified in which targeted correction of the resistance gene had occurred (see Southern blot in Figure 4). From Offringa, R., Van den Elzen, P.J.M., and Hooykaas, P.J.J.,* Transgenic Res., *1, 114, 1992. By permission of Chapman & Hall.*

C. Problems encountered in gene targeting experiments

The design of the different gene targeting experiments described above allows direct selection for gene targeting events since this will restore the activity of a selectable marker gene. In the experiments of Paszkowski et al.[44] and Lee et al.,[73] selection was very tight. A product of targeted recombination was detected in most of the selected calli. In contrast, the experiments of Offringa et al.[72] and Halfter et al.[69] showed a high background of selected but nonrecombinant calli. In both cases, the mutation in the selectable marker gene at the targeting construct was restored rather frequently after integration into the genome, probably by fusion to a plant gene. This is consistent with experiments that show that T-DNAs carrying promoterless marker genes tag plant promoters in 20 to 50% of the transformants.[77–79] The data suggest that T-DNA is inserted predominantly in actively transcribed regions of the ge-

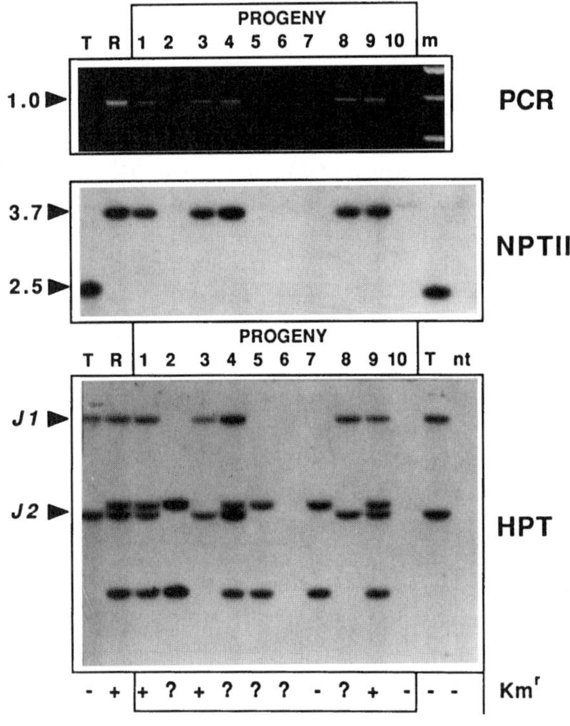

Figure 4. Progeny analysis of the recombinant line R. Plant line R was back-crossed with a non-transgenic tobacco line and the seeds obtained were germinated. The kanamycin resistance phenotype was found to segregate in a 1:1 ratio. The segregation of the targeted locus was analyzed in ten seedlings by PCR with primers 1 and 2 (Figure 3). Furthermore, a blot with HindIII-digested genomic DNA was hybridized with either the NPTII or the HPT probe. As expected, the HPT hybridizing fragments J1 and J2, which were already present in the target line and represent the T-DNA border fragments, co-segregated with the 3.7 kbp (restored) NPTII fragment in the PCR-positive seedlings. The additional HPT hybridizing fragments, which represent the insertion of truncated copies of the repair construct, co-segregated with each other, but not with the restored NPTII fragment. This shows that no alterations occurred at the ends of the target locus during the repair process. Seedlings 1, 3, 7, 9, and 10 were back-crossed again, and seeds were tested for resistance to kanamycin. The PCR-positive seedlings 1, 3, and 9 of the first generation gave rise to kanamycin-resistant progeny in the second generation, confirming the stable functional restoration of one of the defective NPTII genes at the target locus. (Kmr: + = resistant progeny; – = no resistant progeny; ? = not tested). From Offringa, R., Van den Elzen, P.J.M., and Hooykaas, P.J.J., Transgenic Res., *1, 114, 1992. By permission of Chapman & Hall.*

nome. The tagging efficiency with a promoterless marker gene seems to be higher in plants than in mammalian cells, where a splice acceptor site has to be introduced to obtain similar tagging efficiencies.[80] The latter indicates that in many cases, tagging in mammalian cells is directed to the introns. This may be related to the fact that plant genes are, in general, smaller in size and have fewer, generally smaller introns. The high tagging efficiency in plant cells is an important reason for the high rate of activation of defective selectable marker genes that lack a promoter and/or terminator region and are introduced into plant cells via a targeting vector.

In the experiments of Offringa et al.,[72] PCR was used to distinguish the recombinant calli from the background of nonrecombinant ones. The product of homologous recombination (i.e., an intact Kmr gene) was detected in three calli, whereas actual molecular proof for alteration/restoration of the genomic copy was only obtained for one of them. In the two other calli targeted, recombination most likely had ocurred as well. However, since it was not possible to distinguish the recombination products from transformants that may have received a contaminating control construct, these recombinants were not studied in detail.

A marker deletion was introduced into the repair construct so that in subsequent experiments, the product of recombination would be unique (not created before *in vitro*) and distinguishable from the intact Kmr gene at a putative contaminating control construct. To reduce the formation of plant gene/*nptII* fusions one of the T-DNA genes (iaaH) was introduced as a piece of nonhomologous DNA between the T-DNA border repeat and the nptII coding region (Figure 5). The modified repair construct pSDM321 was transformed to protoplasts of the previously used target plant line T[72]. The presence of the iaaH gene resulted in an 8- to 20-fold reduction of the relative number of kanamycin-resistant calli obtained. This is similar to the reduction observed by Herman et al.[78] after introduction of a large piece of DNA upstream of a promoterless selectable marker gene in a T-DNA construct. Two recombinant lines were obtained in which the recombination product was detected by PCR and Southern blot analysis. Further molecular and genetic analysis of these two lines indicated that the target locus had been left unchanged and that the defective Kmr gene at the incoming construct had been repaired using the target locus as template. The corrected construct was inserted at another location in the plant genome.[81,82] The products of recombination could not easily be explained via the DSBR model, which has been proposed for correction of a gapped plasmid by a homologous chromosomal locus in *Saccharomyces cerevisiae*[83] and in mammalian cells.[84,85] Instead, they fitted the model that was first proposed by Adair and co-workers[86] for homologous recombination events observed in mammalian cells, where sequences beyond the region of homology were copied from the target locus to the incoming construct. This end extension repair model involves the invasion of a 3′ 0H strand of the acceptor molecule into the homologous donor duplex and subsequent elonga-

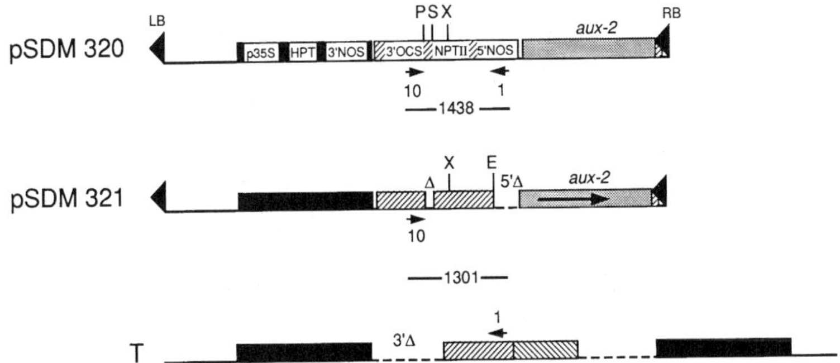

Figure 5. *The T-DNA constructs pSDM320 and pSDM321 and the artificial target locus in plant line T. The arrow in aux-2 (iaaH) indicates the direction of transcription. Abbreviations: p35S = promoter region of the CaMV 35S transcript; HPT or NPTII = region encoding hygromycin or neomycin phosphotransferase, respectively; 5'NOS or 3'NOS = promoter or transcription termination area of the nopaline synthase gene; 3'0CS = transcription termination area of the octopine synthase gene; RB and LB = right and left T-DNA border repeat; X = XhoII; E = EcoRI; S = SmaI; P = PstI.*

tion using the donor DNA as template. Repair of an introduced T-DNA construct in plant cells was also reported by Gal and co-workers,[87] who observed intermolecular recombination in *Brassica napus* between a T-DNA construct harboring an incomplete genome of the cauliflower mosaic virus and a target locus containing the missing sequence with overlapping regions of homology on both sides. In this case, the origin of at least some of the functional virus copies produced could be explained by the DSBR model.

In two of the initial gene targeting experiments in plants, the product of targeted homologous recombination was detected. However, the formation of this product via end extension repair could not be ruled out.[44,73] For future gene targeting experiments, it will be essential that the experimental design allows clear distinction between actual gene targeting and repair of the incoming construct.

D. The fidelity of targeted recombination in plant cells

The accuracy of the gene targeting process can be very high in mammalian cells.[88,89] In order to address this question for plant cells, we performed sequence analysis on the recombination products from the three true recombinant lines described above. No mutations were detected in a 1301-bp PCR fragment amplified from each recombinant Kmr gene (see Figure 5), suggesting that the process also occurs with high accuracy in plant cells.[90] However, since the nptII coding region comprises most of the sequenced fragment, there could have been selection for nonmutated recombination products.

To avoid preselection for perfect targeting events, Paszkowski and co-workers[44] inserted a 1064-bp intron into the nptII coding region. First they proved that the intron was efficiently spliced out and that expression of the gene in plant cells conferred resistance to kanamycin.[91] A transgenic *Nicotiana tabacum* line harboring a construct that contained the intron and the 3′ end of the Kmr gene was used as target line, and protoplasts of this line were transformed via PEG treatment with a construct containing the 5′ part of the gene including the intron. Five recombinant lines were obtained, and sequence analysis of the intron part of three of the recombinant genes revealed only three base-pair substitutions in total.[92] This provides further evidence that gene targeting occurs at high fidelity in plant cells.

More recent results obtained in our group indicate that gene targeting in plants does not always result in perfect events. Using the modified targeting vector pSDM321[82] (Figure 5), five gene targeting events were obtained, four of which coincided with extra rearrangements at the target locus.[93] A similar observation was made in mouse embryonic stem (ES) cells after transformation with a targeting vector having a total homology of 6.8 kbp, but with the homology at one side of the inserted Kmr gene restricted to 0.7 kbp.[84] When the homology at this side was enlarged to 2.7 kbp, all gene targeting events obtained were perfect. The total homology in our T-DNA construct was 3.6 kbp, but at one end, the homology was limited to 0.6 kbp.[82] These data indicate that in both mammalian and plant cells, there is a minimal homology requirement for accurate gene replacement.

V. Towards gene targeting of non-selectable genes

As mentioned above, the initial gene targeting experiments in plant cells were designed such that targeted homologous recombination would lead to the formation of an active selectable marker gene, thus providing a direct selection for gene targeting events. For a number of plant genes, such a selection is available either by gene inactivation (*aprt*, adenine phosphoribosyltransferase,[94] *Nia*, nitrate reductase,[95] *PATl*, phosphoribosylanthranilate transferase[96]) or the transfer of a specific base-pair substitution (*Sur* or *csr*, acetolactate synthase[75]). For the majority of plant genes, however, such a selection system is not available. Moreover, several selection systems used in mammalian cells have proved difficult to apply to plant systems. For example, a specific point mutation in the AtrpII gene encoding the largest subunit of RNA polymerase II renders the protein resistant to the mushroom toxin α–amanitin, which specifically inhibits this subunit.[97] In mouse ES cells, this selection has been used successfully to obtain targeted transfer of the point mutation to the chromosomal locus.[98] Selection experiments with *Arabidopsis* protoplasts performed in our lab revealed, however, that at least a 100-fold higher concentration of α–amanitin was needed for effective selection than in mouse ES cells, and thus, the selection protocol proved to be impractical for large scale gene targeting experiments.[99]

In the relatively short history of gene targeting in mammalian cells, several strategies have been developed that allow the selection of gene targeting events from large numbers of nontargeted clones (see Chapter 1). A discussion of the use and/or usefulness of some of these approaches for gene targeting in plant cells follows.

A. Positive selection and PCR analysis

A general approach to achieve targeted disruption of a gene is to use a targeting vector in which thc coding region of the target gene is disrupted by a functional selectable marker gene, select transformants, and screen for targeted insertion using PCR. In mammalian systems, PCR analysis has proven to allow the reliable detection of recombinant lines in pools of 20 to 50 transformed clones.[100–102] For plant cells, it appears to be possible to detect a recombination event by performing PCR on pooled DNA from 25 transformed lines.[93]

A more specific enrichment for gene targeting events has been obtained by constructing a targeting vector that has a translational fusion between the coding region of the selectable marker gene and that of the target gene. This approach should provide a selective advantage when the promoter and/or terminator region of the target gene is omitted. Such a strategy resulted in a 25- to 100-fold enrichment for gene targeting in mammalian cells.[103–105] Experiments were performed in our group with tobacco protoplasts to evaluate the use of this strategy for targeting the rbcS gene, a nuclear gene coding for the small subunit of ribulose-biphosphate carboxylase.[81,106] Our results showed that nonhomologous integration of the promoterless transgene resulted in its activation at a surprisingly high frequency. This abolishes the selective advantage of using a promoterless translational fusion construct for gene targeting in plant cells.

B. Positive/negative selection: selection markers for plant cells

The most successful and commonly used strategy to enrich for gene targeting in mammalian cells is positive/negative selection[107] (see Chapter 1). A prerequisite for using this strategy in plant systems is the development of negative selectable marker genes for plant cells. Although a number of well-defined and robust positive selection systems are available, negative selection systems for plant cells are not well developed.

One of the first genes that was used as negative selectable marker in plant cells was the *iaaH* gene encoding an indole-3-acetamide hydrolase, which converts the low activity auxin naphthalene acetamide (NAM) into the active auxin naphthalene acetic acid (NAA).[108,109] This phytohormone is toxic to plant cells at high concentrations, and thus, the addition of high concentrations of NAM to the culture medium will kill cells that are able to efficiently convert it into an active auxin. NAM selection has been reported to work both on protoplast level[110] and on seedling level.[111] The *iaaH* gene, referred to as *aux-2* in Figure 5, was tested for use in gene targeting experiments in our group. In one experiment a three- to four-fold enrichment was obtained and a targeted homologous recombination event was selected.[81] The efficiency of targeted recombination (1 in 7×10^4 transformants) was not significantly reduced by the presence of the nonhomologous iaaH gene on the end of the replacement-type targeting vector.[81] This is similar to the observations in mammalian cells 10[7]. In other experiments, however, the non-cell-autonomous behaviour of iaaH resulted in killing of cells that did not contain the gene. This clearly indicates that only cell-autonomous markers should be considered for use as negative markers in gene targeting experiments.

Several negative selection markers that have successfully been used to enrich for gene targeting in mammalian cells are now being tested in plant cells. One of them, the gene from Corynebacterium diphtheriae encoding the A fragment of the diphtheria toxin (DT-A), has been used as a nonconditional negative selectable marker to disrupt the c-fyn gene in mouse ES cells.[112] DT-A blocks protein synthesis by ADP-ribosylating elongation factor 2 and has been shown to be very toxic to plant cells[93,113,114] and to work in a cell-autonomous way.[113] For a nonconditional marker like DT-A to be useful, expression should only occur after integration, since transient expression before integration may kill potential recombinants. This is important especially for DT-A, since there is evidence that one molecule per cell is lethal.[115] Yagi and co-workers[112] overcame this problem in mouse ES cells by omitting the transcription termination signals from the marker gene. Preliminary results in tobacco protoplasts suggest that this may not be sufficient to eliminate transient expression in plant cells.[93] An alternative approach may be to convert the gene into a conditional marker by putting it under the control of an inducible promoter.[116] Because of the high toxicity of DT-A, a tight control of gene expression will be essential. In this respect, the gene encoding the attenuated form of DT-A may be a better selectable marker.[117,118] Clearly the use of conditional marker genes is to be preferred over the nonconditional ones. A conditional selectable marker that is commonly used now in mammalian systems is the thymidine kinase gene of herpes simplex virus (HSV-*tk*), which causes sensitivity to the nucleoside analogue gancyclovir.[107] HSV-*tk* expression in *Arabidopsis* plants under control of a plant promoter conferred sensitivity to gancyclovir;[119] however, after prolonged selection, transformed tissues developed slow-growing calli.[120] Although gancyclovir selection is leaky at the callus level, preliminary data suggest that gancyclovir selectively inhibits shoot regeneration from root tissue of an *Arabidopsis* line harboring the HSV-*tk* gene.[121]

Recently, the coding region of the nitrate reductase (Nia) placed under control of the constitutive CaMV 35S promoter was used as negative selectable marker in the tobacco species *Nicotiana plumbagenifolia*. Cells overexpressing

Nia from this gene construct convert chlorate added to the medium to chlorite, which is toxic to plant cells. Selection against nontransformed cells is avoided by growing cells on nitrate deficient medium, which results in repression of the expression of the endogenous Nia gene. Chlorate selection worked efficiently on the seedling level, but the chlorate dose that was needed for efficient counter selection in cell cultures reduced the survival of nontransformed cells to 40%.[122]

Perhaps the most promising conditional selectable marker for plant cells is the cytosine-deaminase gene (*codA*) from *Escherichia coli.* In bacteria, cytosine deaminase (CD) mediates the conversion of cytosine into uracil.[123] This pathway is absent in plants and mammals where cytosine appears biochemically inert.[124,125] The CD enzyme also deaminates 5-fluorocytosine (5-FC) into 5-fluorouracil (5-FU), a precursor of 5-fluoro-dUMP that irreversibly inhibits thymidylate synthase activity and is lethal to the cell. Murine fibroblasts transfected with the codA gene were effectively eliminated by 5-FC selection without nonspecific killing of mixed wild-type cells.[126] Negative selection with 5-FC was also effective with callus tissues or seedlings of the plant species *Lotus japonicus.*[127] Whether the negative effect is cell autonomous and whether 5-FC selection will also work with other plant species is now subject to further investigations.

In addition to the specialized negative selectable marker genes, two research groups have tested the use of antisense constructs that silence the positive selectable marker gene present in the targeting construct as negative markers. Xiang and Guerra[120] found that *nptll* gene expression in tobacco cells can be repressed when the antisense gene is present in the same construct. The Kmr phenotype is overcome by antisense RNA only if sufficient kanamycin is supplied to the selection medium. Halfter[71] used a similar antisense approach to enrich for targeted disruption of the G-protein encoding *YPT* gene from *Arabidopsis thaliana* by a sense *hpt* gene. With the construct harboring the antisense gene, the number of hygromycin resistant lines obtained was only reduced two- to three-fold compared to a control construct that lacked the antisense gene. The results suggest that silencing is never complete and that a strong reduction of transformation can only be obtained by applying high selection pressure. More importantly, the presence of the internal homology in the targeting construct may lead to intramolecular recombination and, as such, may interfere with the efficiency of gene targeting.

VI. Factors that may affect gene targeting in plants

A. Method of DNA transfer

The published data on gene targeting in plants (see above) do not allow a direct comparison between different methods of DNA transfer, since different targeting constructs and different target lines were used. However, in our group, a similar targeting vector was introduced into protoplasts of the same target lines either via electroporation or via *Agrobacterium*-mediated DNA transfer. A comparison of the results (Table 1) shows that with the latter method, gene targeting was reproducibly obtained,[72,81,82,93] whereas no targeting events were detected after electroporation.[106] These results indicate that the *Agrobacterium* vector system is at least as efficient for gene targeting in plants as direct DNA transfer. Apparently the mechanism by which T-DNA is transferred to the plant cell nucleus does not interfere with its interaction with homologous DNA sequences in the nucleus.

Agrobacterium-mediated DNA transfer results in a relatively high proportion of single copy transformants.[34,79,128–130] This is an advantage when negative selection is applied to enrich for gene targeting events, since gene targeting events coinciding with extra nonhomologous insertions will be lost. The absolute targeting frequency in a mouse target cell line was not reduced when a lower number of targeting vector DNA copies was introduced. Instead, the specific reduction of the number of nonhomologous integration events resulted in a higher relative frequency of gene targeting.[84]

A disadvantage of *Agrobacterium*-mediated DNA transfer is that the T-strand, which is provided by *Agrobacterium* as targeting vector, cannot be manipulated very easily. For naked DNA transfer, any unique restriction site in the plasmid can simply be used as site of linearization. In this way, a plasmid can be converted into either a replacement- or insertion-type vector (see Chapter 1). The T-strand is a natural replacement vector, but it can be converted into an insertion vector depending on the way the targeting construct is cloned between the T-DNA borders, since the border repeats mimic the endonucleolytic cleavage sites in a plasmid. Wild type T-DNA border fragments introduce some nonhomology at the termini of transferred DNA, but by using synthetic T-DNA border repeats,[131] this nonhomology can be minimized to only a few base-pair. Nonhomology at the ends of a replacement-type vector does not influence the efficiency of gene targeting in mammalian cells.[107,132] Nonhomology at the ends of an insertion-type vector does reduce the targeting frequency in mammalian cells,[132,133] although this effect is not observed when the nonhomology is limited to a few base-pairs.[133]

In some of the gene targeting experiments in plants using direct DNA transfer methods, carrier DNA was added[44,69] because it has been found to increase the efficiency of transformation.[11,134,135] However, carrier DNA and plasmid DNA are known to form concatemers that are integrated at a limited number of chromosomal sites.[136,137] Carrier DNA may protect the plasmid DNA and, thus, the homologous region, from degradation by nucleases, as was suggested by experiments concerning extrachromosomal recombination in plants.[51] However, its presence may inhibit homologous recombination by concealing the regions of homology, or by diluting enzymes that are involved in recombination, as reported in yeast.[138] Carrier DNA is normally not used in gene targeting experiments with mammalian systems.

***Table 1.** T-DNA transfer vs. electroporation*

Method	Exp. n°[a].	Protoplasts	Trans-formants	Targeted recombination	RRF[b]	Ref.
T-DNA transfer	I	1.5×10^7	1.2×10^5	1	8.3×10^{-6}	72
	II	3.6×10^7	7.0×10^5	1	1.4×10^{-6}	82
	III	9.3×10^6	7.0×10^4	1	1.4×10^{-5}	81
	IV	4.7×10^7	1.6×10^6	6	3.8×10^{-6}	93
Electroporation	V	1.4×10^8	2.2×10^5	0	$< 4.5 \times 10^{-6}$	106
	VI	9.0×10^7	1.4×10^5	0	$< 7.1 \times 10^{-6}$	106
	VII	9.9×10^7	5.6×10^4	0	$< 1.8 \times 10^{-5}$	106

[a] Experiments I, V, and Vl were performed with the same targeting vector and target line and are most comparable. However, in experiment VI the targeting vector was circular ssDNA, whereas in the other electroporation experiments, linear dsDNA was used. Experiments II, III, and IV were performed with the modified construct (See Figure 5, contains the *iaaH* gene). The recombinant from experiment III was obtained after NAM (negative) selection. Experiment IV is a summary of the data obtained with six new target lines. One of the new target lines from which two recombinants were obtained was used in experiment VII.

[b] RRF = relative recombination frequency: the number of targeted recombinants (both actual gene targeting and correction of incoming construct) divided by the number of transformed surviving protoplasts.

B. Position of target locus in genome and genome size

The essential first step in gene targeting is that the targeting construct finds the homologous target in the genome. This process is likely to be hampered by the location of the target locus in densely packed chromatin structures[102,139] or by a large genome size.

Halfter and co-workers[69] performed gene targeting experiments in *Arabidopsis*, whereas the other plant groups worked with tobacco. In spite of the small genome size of *Arabidopsis thaliana* (108 bp, which is only 1/20th of the tobacco genome) and the relatively low abundance of repetitive sequences,[140,141] the frequencies of targeted recombination were not significantly higher in *Arabidopsis*. This suggests that, as in mammalian cells, searching for homology is not the rate-limiting step in plants.[142]

Both Paszkowski et al.[44] and Halfter et al.[69] used different target lines as acceptor for transformation with the targeting construct. Targeted recombination was only obtained in the minority of the lines, suggesting that the efficiency of gene targeting is subject to line-to-line variations, which may be determined by specific chromosomal sequences in the vicinity of the target locus. A similar conclusion was reached after gene targeting experiments in mammalian cells.[85,143] However, some of the lines used harbored more target inserts than others, and therefore, a direct comparison is difficult.

We addressed the question by constructing different target lines that harbor only one intact copy of a target T-DNA.[128] Five of these lines were used for transformation with the targeting construct pSDM321 (Figure 5), and although only one or two recombination events were obtained per line, the number of transformants needed to obtain one recombination event was comparable between lines.[93] Thus, in our experiments, no clear line-to-line variation was observed. However, it should be noted that artificial T-DNA target loci were used in this experiment. These may themselves be located in accessible regions of the chromatin, since T-DNA integration may only occur in such regions. The situation may be quite different for endogenous loci.

C. DNA sequence homology

A clear positive correlation has been found in mouse ES cells between the amount of homology present in the targeting construct and the efficiency of targeting.[144,145,146] Deng and Capecchi[146] calculated the absolute targeting frequency (ATF, being the number of targeted recombination events per number of cells used for transfection) for different replacement-type and insertion-type vectors and found an exponential relationship between the length of homology and the ATF. The recombination system appears to be saturated with respect to this parameter at approximately 14 kbp of homology.

Only replacement-type vectors have been used in the four initial targeting experiments in plant cells. The amount of homology used and the ATF are presented in Table 2. The ATF was calculated by dividing the number of targeted recombinants by the number of surviving protoplasts from target lines giving rise to recombinants. This provided an estimation of the maximal ATF value for the given region of homology. For the experiments of Offringa et al.[72] and Halfter et al.,[69] only those clones for which targeted recombination was proven at the molecular level have been taken into account. No correlation can be observed between the length of homology and the ATF in plant cells, but this may be explained by the large differences in test systems and the relatively small regions of homology in the targeting constructs. However, the ATF in plant cells is comparable to the ATF obtained by Deng and Capecchi[146] in mouse ES cells with a replacement vector having 2.8 kbp homology. This suggests that, in plant cells, the ATF may also be increased considerably by using larger stretches of homology.

As described above, Lee et al.[73] used the endogenous acetolactate synthase (ALS) gene combined with the *Agrobacterium* vector system for gene targeting experiments in tobacco. Considering the fact that there are two genetically unlinked ALS loci in tobacco (i.e., SurA and SurB), it was remarkable that targeted recombination was only obtained with the fully homologous SurB locus and not with the less homologous SurA locus. This finding is consistent with the observation in mouse ES cells[147] that there is a strong bias in gene targeting towards the locus with the highest homology. It also indicates that the

Table 2. *Lengths of homologous overlaps and absolute targeting frequencies (ATF) observed in plant and mouse ES cells*

First author (ref.)	Homology (kbp)	Cells surviving	Targeted recombination	ATF
Paszkowski[44]	0.4	$7.7 \times 10^{6(a)}$	8	1.0×10^{-6}
Lee[73]	1.9	1.1×10^{6}	3	2.8×10^{-6}
Offringa[72,81,82] and Risseeuw[93]	3.6	1.1×10^{7}	$9^{(c)}$	0.8×10^{-6}
Halfter[69–71]	1.1	$3.5 \times 10^{5(b)}$	2	5.7×10^{-6}
Deng and Capecchi[146]	2.8	—	—	2×10^{-6}

[a,b] Obtained by multiplying the number of treated cells with a plating efficiency of (a) 10%[134] and (b) 0.5%.[135]

[c] A summary of the experiments of Offringa and Risseeuw.

observation that much higher gene targeting efficiencies are obtained in mammalian cells if the targeting construct is prepared from an isogenic compared to a nonisogenic DNA source[146,148] may apply to plant cells as well.

D. Cell type

Most of the successful gene targeting experiments in mammalian systems have been performed using mouse ES cells. Thus, it is possible that the use of embryonic plant cells may be advantageous for gene targeting. Several protocols have been developed to obtain transformants from embryogenic plant cell cultures[149,150] or plant embryos.[16,151,152] Although these systems are not yet comparable with gene targeting experiments, after further optimization, they may be used to determine if cell type has an influence on targeted recombination. Cell type can influence gene targeting in animal cells as shown by Buerstedde and Takeda.[153] Their work indicated that the frequencies of gene targeting were significantly higher in chicken B cell lines, in which the V(D)J recombination system of the immunoglobulin genes is active, than in non-B cell lines. The search for recombination-efficient cell types in plants has not been started yet.

E. Enhancement of the homologous recombination pathway in plants

In previous chapters (1 and 2), several treatments have been described that may enhance the efficiency of targeted recombination. Pre-treatment of cells or tissues with UV light, X-rays, or other DNA-damaging agents is known to induce genes that are involved in DNA repair and recombination. Moreover, these treatments introduce nicks and double strand breaks into the chromosomal DNA, which may be initiation points for recombination. Although gamma- and X-irradiation increase the frequency of intrachromosomal homologous recombination in tobacco cells,[62,63] these treatments enhance nonhomologous recombination as well[155,156] and, more importantly, affect cell viability[62,63] and may introduce additional unwanted mutations. More promising is the treatment with mitomycin-C, a compound that induces crosslinking between DNA strands, which gives a large increase in the frequency of intrachromosomal homologous recombination in somatic plant cells without reducing cell viability.[62]

Recently a maize inbred line was described in which mitotic homologous recombination between duplicated zein genes was detected at high frequency.[154] Thus, it may be possible to use plant varieties or mutants that show an increased frequency of homologous recombination.

For the specific manipulation of the recombination pathway towards gene targeting, it is essential to gain more knowledge of enzymes involved in DNA recombination and/or repair. Although this line of research is still in its infancy for plants, some of the enzymatic activites involved in mismatch repair[157,158] and UV-induced endonuclease activity[159] have been identified in tobacco cells. A type-I topoisomerase has been purified from tobacco[160] and from broccoli.[161] The cDNA of the topoisomerase I gene in *Arabidopsis thaliana* has been cloned.[162] Recently, several cDNAs have been cloned from *Arabidopsis thaliana,* either by homology to the cyanobacterial recA protein[163] or by complementation of repair-deficient *E.coli* strains.[164,165] An interesting observation was that the proteins encoded by these cDNAs all have an N-terminal transit peptide for transport into the stroma of the chloroplast, where exogenous DNA is predominantly integrated via homologous recombination.[35]

VII. Future perspectives for the site-directed mutagenesis of the plant genome

The final and most important question is whether gene targeting in plant cells can ever be optimized such that it will be as efficient as in certain mammalian cell systems. Although the data concerning ECR, ICR, and gene targeting indicate many similarities between the processes of homologous recombination in plant and mammalian cells, there may be a crucial difference that has not yet been discovered. A permanent low efficiency of gene targeting in plant cells would mean that the gene targeting technology would be restricted to model plant species for which efficient transformation systems have been developed. For most crop species, the maximal number of transgenic lines that can routinely be obtained is limited, varying from one to a few hundred. It may, however, be possible to develop alternative methods to obtain targeted disruption or replacement of a gene.

One alternative may be to use vectors that replicate in the plant cell to deliver the targeting construct. The geminiviruses have drawn attention as replication vectors for the introduction of foreign DNA into plant cells. They have a mono- or bipartite genome of ssDNA, and a group of geminiviruses can infect a wide range of monocotyledonous and dicotyledonous plants.[166] Several groups have shown that recombinant geminiviruses in which the coat protein gene (which is not essential for systemic spread throughout the plant) has been replaced by a bacterial reporter gene can be replicated to a high copy number per cell and spread throughout the plant.[166–168] The results of Gal and co-workers[87] indicate that the dsDNA caulimovirus CaMV can aquire sequences from the plant genome via homologous recombination. Whether this will also occur in the reverse direction is not known. Moreover, homologous or nonhomologous interactions between plant nuclear DNA and geminiviruses have never been reported.

Another option is to introduce the targeting vector into an unlinked ectopic site of the plant genome and to select for homologous recombination between this site and the target locus in the progeny of the transgenic lines. Once recombination with the target locus has been obtained, the ectopic copy can be removed by crossing. Whether homologous recombination between a specific locus and a ectopic copy will occur at a detectable efficiency in plants is still under investigation. From earlier studies in yeast and mammalian cells, it is known that a DSB in the acceptor DNA molecule will enhance the transfer of information from the donor DNA via gap repair.[169,170] Moreover, in *Drosophila melanogaster,* it was initially found that excision of the transposable P element leaves a DSB at the chromosome, which can use either the homologous chromosome or homology at an ectopic locus as a template for gap repair.[171,172] Similar data were obtained for the excision of the transposable Tcl element in the nematode *Caenorhabditis elegans*.[173] In both organisms, transposition can be used to enhance transfer of DNA sequence information from ectopically located homologous DNA to the target locus. In the future, this system may also be used in plants, provided that plant transposable elements introduce a DSB during transposition and that plant lines are available containing transposon insertions mapping to the desired genomic position.

As the crucial gene targeting experiments with optimized targeting vectors containing large regions of homology and well-developed selection systems are presently being performed, within a few years it will become clear whether or not the gene targeting technology will become generally applicable in plants. Although we are still confident that this will be the case, its use may remain restricted to reverse genetics research in model plant species. The development of alternative methods such as the ones discussed above may allow the site directed mutagenesis of the genome of a wider range of plant species.

Acknowledgments

We would like to thank Jerzy Paszkowski, Laszlo Marton, Stephan Ohl, and Eddy Riseeuw for sharing their unpublished data and Roger Morton for critical reading of the manuscript.

References

1. De Block, M., Botterman, J., Vandewiele, M., Dockx, J., Thoen, C., Gosselé, V., Mowa, N.R., Thompson, C., Van Montagu, M., and Leemans, J., Engineering herbicide resistance in plants by expression of a detoxifying enzyme., *EMBO J.*, 6, 2513, 1987.

2. Gray, J., Picton, S., Shabbeer, J., Schuch, W., and Grierson, D., Molecular biology of fruit ripening and its manipulation with antisense genes, *Plant Mol. Biol.*, 19, 69, 1992.

3. Knutzon, D.S., Thompson, G.A., Radke, S.E., Johnson, W.B., Knauf, V.C., and Kridl, J.C., Modification of *Brassica* seed oil by antisense expression of a stearoyl-acyl carrier protein desaturase gene, *Proc. Natl. Acad. Sci. U.S.A.*, 89, 2624, 1992.

4. Meeusen, R.L., and Warren, G., Insect control with genetically engineered crops, *Annu. Rev. Entomol.*, 34, 373, 1989.

5. Anzai, H., Yoneyama, K., and Yamaguchi, I., Transgenic tobacco resistant to a bacterial disease by the detoxification of a pathogenic toxin, *Mol. Gen. Genet.*, 219, 492, 1989.

6. Abel, P.P., Nelson, R.S., De, B., Hoffmann, N., Rogers, S.G., Fraley, R.T., and Beachy, R.N., Delay of disease development in transgenic plants that express the tobacco mosaic virus coat protein gene, *Science*, 232, 738, 1986.

7. Cornelissen, B.J.C., and Melchers, L.S., Strategies for control of fungal diseases with transgenic plants, *Plant Physiol.*, 101, 709, 1993.

8. Roth, D.B., and Wilson, J.H., Illegitimate recombination in mammalian cells, in *Genetic Recombination*, Kucherlapati, R., and Smith, G.R., Eds., American Society for Microbiology, Washington, DC, 1988, 621.

9. Gheysen, G., Villarroel, R., and Van Montagu, M., Illegitimate recombination in plants: a model for T-DNA integration, *Genes Dev.*, 5, 287, 1991.

10. Mayerhofer, R., Koncz-Kalman, Z., Nawrath, C., Bakkeren, G., Crameri, A., Angelis, K., Redei, G.P., Schell, J., Hohn, B., and Koncz, C., T-DNA integration: a mode of illegitimate recombination in plants, *EMBO J.*, 10, 697, 1991.

11. Krens, F.A., Molendijk, L., Wullems, G.J., and Schilperoort, R.A., *In vitro* transformation of plant protoplasts with Ti-plasmid DNA, *Nature*, 296, 72, 1982.

12. Shillito, R.D., Saul, M.W., Paszkowski, J., Muller, M., and Potrykus, I., High efficiency direct gene transfer to plants, *Bio/technol.*, 3, 1099, 1985.

13. Deshayes, A., Herrera-Estrella, L., and Caboche, M., Liposome-mediated transformation of tobacco mesophyll protoplasts by an *Escherichia coli* plasmid, *EMBO J.*, 4, 2731, 1985.

14. Crossway, A., Oakes, J.V., Irvine, J.M., Ward, B., Knauf, V.C., and Shewmaker, C.K., Integration of foreign DNA following microinjection of tobacco mesophyll protoplasts, *Mol. Gen. Genet.*, 202, 179, 1986.

15. Klein, T.M., Harper, E.C., Svab, Z., Sanford, J.C., Fromm, M.E., and Maliga, P., Stable genetic transformation of intact *Nicotiana* cells by the particle bombardment process, *Proc. Natl. Acad. Sci. U.S.A.*, 85, 8502, 1988.

16. D'Halluin, K., Bonne, E., Bossut, M., De Beuckeleer, M., and Leemans, J., Transgenic maize plants by tissue electroporation, *Plant Cell*, 4, 1495, 1992.

17. Gasser, C.S., and Fraley, R.T., Genetically engineering plants for crop improvement, *Science*, 244, 1293, 1989.

18. Marton, L., Wullems, G.J., Molendijk, L., and Schilperoort, R.A., *In vitro* transformation of cultured cells from *Nicotiana tabacum* by *Agrobacterium tumefaciens*, *Nature*, 277, 129, 1979.

19. Horsch, R.B., Fry, J.E., Hoffmann, N.L., Eichholtz, D., Rogers, S.G., and Fraley, R.T., A simple and general method for transferring genes into plants, *Science*, 227, 1229, 1985.

20. Sheerman, S.A., and Bevan, M.W., A rapid transformation method for *Solanum tuberosum* using binary *Agrobacterium tumefaciens* vectors, *Plant Cell Rep.*, 7, 13, 1988.

21. Valvekens, D., Van Montagu, M., and Van Lijsebettens, M., *Agrobacterium tumefaciens* mediated transformation of *Arabidopsis thaliana* root explants by using kanamycin selection, *Proc. Natl. Acad. Sci. U.S.A.*, 85, 5536, 1988.

22. Hooykaas-Van Slogteren, G.M., Hooykaas, P.J.J., and Schilperoort, R.A., Expression of Ti plasmid genes in monocotyledonous plants infected with *Agrobacterium tumefaciens*, *Nature*, 311, 763, 1984.

23. Hernalsteens, J.-P., Thia-Toong, L., Schell, J., and Van Montagu, M., An *Agrobacterium* transformed cell culture from the monocot *Asparagus officinalis*, *EMBO J.*, 3, 3039, 1984.

24. Hohn, B., Koukolikova-Nicola, Z., Bakkeren, G., and Grimsley, N., *Agrobacterium*-mediated gene transfer to monocots and dicots, *Genome*, 31, 987, 1989.

25. Gould, J., Devey, M., Hasegawa, O., Ulian, E.C., Peterson, G., and Smith, R.H., Transformation of Zea mays L using *Agrobacterium tumefaciens* and the shoot apex, *Plant Physiol.*, 95, 426, 1991.

26. Zambryski, P.C., Chronicles from the *Agrobacterium*-plant cell DNA transfer story, *Annu. Rev. Plant Physiol. Plant Mol. Biol.*, 43, 465, 1992.

27. Hooykaas, P.J.J., and Schilperoort, R.A., *Agrobacterium* and plant genetic engineering, *Plant Mol. Biol.*, 19, 15, 1992.

28. Herrera-Estrella, A., Van Montagu, M., and Wang, K., A bacterial peptide acting as a plant nuclear targeting signal: the amino-terminal portion of *Agrobacterium* VirD2 protein directs a β-galactosidase fusion protein into tobacco nuclei, *Proc. Natl. Acad Sci. U.S.A.*, 87, 9534, 1990.

29. Howard, E.A., Zupan, J.R., Citovsky, V., and Zambryski, P.C., The VirD2 protein of *Agrobacterium tumefaciens* contains a C-terminal bipartite nuclear localization signal: implications for nuclear uptake of DNA in plant cells, *Cell*, 68, 109, 1992.

30. Citovsky, V., Zupan, J., Warnick, D., and Zambriski, P., Nuclear localization of *Agrobacterium* VirE2 protein in plant cells, *Science*, 256, 1802, 1992.

31. Tinland, B., Koukolokova-Nicola, Z., Hall, M.N., and Hohn, B., The T-DNA linked VirD2 protein contains two distinct functional nuclear localization signals, *Proc. Natl. Acad. Sci. U.S.A.*, 89, 7442, 1992.

32. Hain, R., Stabel, P., Czernilofsky, A.P., Steinbiß, H.H., Herrera-Estrella, L., and Schell, J., Uptake, integration, expression and genetic transmission of a selectable chimeric gene by plant protoplasts, *Mol. Gen. Genet.*, 199, 161, 1985.

33. Czernilofsky, A.P., Hain, R., Herrera-Estrella, L., Lörz, H., Goyvaerts, E., Baker, B.J., and Schell, J., Fate of selectable marker DNA integrated into the genome of *Nicotiana tabacum*, *DNA*, 5, 101, 1986.

34. Deroles, S.C., and Gardner, R.C., Expression and inheritance of kanamycin resistance in a large number of transgenic petunias generated by *Agrobacterium*-mediated transformation, *Plant Mol. Biol.,* 11, 355, 1988.

35. Maliga, P., Towards plastid transformation in flowering plants, *Trends Bio/technol.*, 11, 101, 1993.

36. Porter, R.D., Modes of gene transfer in bacteria, in *Genetic Recombination*, Kucherlapati, R. and Smith, G.R., Eds., American Society for Microbiology, Washington, DC, 1988, 1.

37. Hinnen, A., Hicks, J.B., and Fink, G.R., Transformation of yeast, *Proc. Natl. Acad. Sci. U.S.A.*, 75, 1929, 1978.

38. Ten Asbroek, A.L.M.A., Quellette, M., and Borst, P., Targeted insertion of the neomycin phosphotransferase gene into the tubulin gene cluster of *Trypanosoma brucei*, *Nature*, 348, 174, 1990.

39. De Block, M., Schell, J., and Van Montagu, M., Chloroplast transformation by *Agrobacterium tumefaciens*, *EMBO J.*, 4, 1367, 1985.

40. Venkateswarlu, K., and Nazar, R.N., Evidence for T-DNA mediated gene targeting to tobacco chloroplasts, *Bio/technol.*, 9, 1103, 1991.

41. Howell, S.H., Walker, L.L., and Walden, R.M., Rescue of *in vitro* generated mutants of cloned cauliflower virus genome in infected plants, *Nature*, 293, 483, 1981.

42. Lebeurier, G., Hirth, L., Hohn, B., and Hohn, T., *In vivo* recombination of cauliflower mosaic virus DNA, *Proc. Natl. Acad. Sci. U.S.A.*, 79, 2932, 1982.

43. Wirtz, U., Schell, J., and Czernilofsky, A.P., Recombination of selectable marker DNA in *Nicotiana tabacum*, *DNA*, 6, 245, 1987.

44. Paszkowski, J., Baur, B., Bogucki, A., and Potrykus, I., Gene targeting in plants., *EMBO J.*, 7, 4021, 1988.

45. Bollag, R.J., Waldman, A.S., and Liskay, R.M., Homologous recombination in mammalian cells, *Annu. Rev. Genet.*, 23, 199, 1989.

46. Baur, M., Potrykus, I., and Paszkowski, J., Intermolecular homologous recombination in plants, *Mol. Cell. Biol.*, 10, 492, 1990.

47. Bilang, R., Peterhans, A., Bogucki, A., and Paszkowski, J., Single-stranded DNA as a recombination substrate in plants as assessed by stable and transient recombination assays, *Mol. Cell. Biol.*, 12, 329, 1992.

48. De Groot, M.J.A., Offringa, R., Does, M.P., Hooykaas, P.J.J., and Van den Elzen, P.J.M., Mechanisms of intermolecular homologous recombination in plants as studied with single- and double-stranded DNA molecules, *Nucleic Acids Res.*, 20, 2785, 1992.

49. Engels, P., and Meyer, P., Comparison of homologous recombination frequencies in somatic cells of petunia and tobacco suggest two distinct recombination pathways, *Plant J.*, 2, 59, 1992.

50. Puchta, H., and Hohn, B., A transient assay in plant cells reveals a positive correlation between extrachromosomal recombination rates and length of homologous overlap, *Nucleic Acids Res.*, 19, 2693, 1991.

51. Puchta, H., and Hohn, B., The mechanism of extrachromosomal homologous DNA recombination in plant cells, *Mol. Gen. Genet.*, 230, 1, 1991.

52. Puchta, H., Kocher, S., and Hohn, B., Extrachromosomal homologous DNA recombination in plant cells is fast and is not affected by CpG methylation, *Mol. Cell. Biol.*, 12, 3372, 1992.

53. Lyznik, L.A., Mcgee, J.D., Tung, P.Y., Bennetzen, J.L., and Hodges, T.K., Homologous recombination between plasmid DNA molecules in maize protoplasts, *Mol. Gen. Genet.*, 230, 209, 1991.

54. Folger, K.R., Thomas, K.C., and Capecchi, M.R., Efficient correction of mismatched bases in plasmid heteroduplexes injected into cultured mammalian cell nuclei, *Mol. Cell. Biol.*, 5, 70, 1985.

55. Lin, F., Sperle, K., and Sternberg, N., Model for homologous recombination during transfer of DNA into mouse L cells: role for DNA ends in the recombination process, *Mol. Cell. Biol.*, 4, 1020, 1984.

56. Lin, F.L.M., Sperle, K.M., and Sternberg, N.L., Extrachromosomal recombination in mammalian cells as studied with single- and double-stranded DNA substrates, *Mol. Cel. Biol.*, 7, 129, 1987.

57. Lin, F., Sperle, K., and Sternberg, N.L., Intermolecular recombination between DNAs introduced into mouse L cells by a nonconservative pathway that leads to crossover products, *Mol. Cell. Biol.*, 10, 103, 1990.

58. Wake, C.T., Vernaleone, F., and Wilson, J.H., Topological requirements for homologous recombination among DNA molecules transfected into mammalian cells, *Mol. Cell. Biol.*, 5, 2080, 1985.

59. Szostak, J.W., Orr-Weaver, T.L., Rothstein, R.J., and Stahl, F.W., The double-strand-break repair model, *Cell*, 33, 25, 1983.

60. Peterhans, A., Schlüpmann, H., Basse, C., and Paszkowski, J., Intrachromosomal recombination in plants, *EMBO J.*, 9, 3437, 1990.

61. Assaad, F.F., and Signer, E.R., Somatic and germinal recombination of a direct repeat in *Arabidopsis*, *Genetics*, 132, 553, 1992.

62. Lebel, E.G., Masson, J., Bogucki, A., and Paszkowski, J., Stress-induced intrachromosomal recombination in plant somatic cells, *Proc. Natl. Acad. Sci. U.S.A.*, 90, 422, 1993.

63. Tovar, J., and Lichtenstein, C., Somatic and meiotic chromosomal recombination between inverted duplications in transgenic tobacco plants, *Plant Cell*, 4, 319, 1992.

64. Gal, S., Pisan, B., Hohn, T., Grimsley, N., and Hohn, B., Genomic homologous recombination in plants, *EMBO J.*, 10, 1571, 1991.

65. Swoboda, P., Hohn, B., and Gal, S., Somatic homologous recombination in plants: the recombination frequency is dependent on the allelic state of recombining sequences and may be influenced by genomic positions, *Mol. Gen. Genet.*, 237, 33, 1993.

66. Swoboda, P., unpublished data, 1993.

67. Wang, Y., Maher, V.M., Liskay, R.M., and McCormick, J.J., Carcinogens can induce homologous recombination between duplicated chromosomal sequences in mouse L cells, *Mol. Cell. Biol.*, 8, 196, 1988.

68. Waldman, A.S., and Liskay, R.M., Differential effects of base-pair mismatch on intrachromosomal versus extrachromosomal recombination in mouse cells, *Proc. Natl. Acad. Sci. U.S.A.*, 84, 5340, 1987.

69. Halfter, U., Morris, P.-C., and Willmitzer, L., Gene targeting in *Arabidopsis thaliana*, *Mol. Gen. Genet.*, 231, 186, 1992.

70. Halfter, U., Studien zur homologen rekombination in *Arabidopsis thaliana* (L.) Heynh., Ph.D. Thesis, IGF Berlin GmbH, Berlin, 1990.

71. Morris, P.-C., unpublished data, 1993.

72. Offringa, R., De Groot, M.J.A., Haagsman, H.J., Does, M.P., Van den Elzen, P.J., and Hooykaas, P.J.J., Extrachromosomal homologous recombination and gene targeting in plant cells after *Agrobacterium*-mediated transformation, *EMBO J.*, 9, 3077, 1990.

73. Lee, K.Y., Lund, P., Lowe, K., and Dunsmuir, P., Homologous recombination in plant cells after *Agrobacterium*-mediated transformation, *Plant Cell*, 2, 415, 1990.

74. LaRossa, R.A., and Schloss, J.V., The sulfonylurea herbicide sulfmeturon methyl is an extreme potent and selective inhibitor of acetolactate synthase in *Salmonella typhimurium*, *J. Biol. Chem.*, 259, 8753, 1984.

75. Lee, K.Y., Townsend, J., Tepperman, J., Black, M., Chui, C.F., Mazur, B., Dunsmuir, P., and Bedbrook, J., The molecular basis of sulfonylurea herbicide resistance in tobacco, *EMBO J.*, 7, 1241, 1988.

76. Chaleff, R.S., Sebastion, S.A., Creason, G.L., Mazur, B.J., Falco, S.C., Ray, T.B., Mauvais, C.J., and Yadav, N.B., Developing plant varieties resistant to sulfonylurea herbicides, in *Molecular Strategies for Crop Protection: UCLA Symposium on Molecular and Cellular Biology*, Arntzen, C. and Ryan, C., Eds., John Wiley & Sons, New York, 1986, 415.

77. Koncz, C., Martini, N., Mayerhofer, R., Koncz-Kalman, Z., Körber, H., Redei, G.P., and Schell, J., High-frequency T-DNA-mediated gene tagging in plants, *Proc. Natl. Acad. Sci. U.S.A.*, 86, 8467, 1989.

78. Herman, L., Jacobs, A., Van Montagu, M., and Depicker, A., Plant chromosome/marker gene fusion assay for study of normal and truncated T-DNA integration events, *Mol. Gen. Genet.*, 224, 248, 1990.

79. Kertbundit, S., de Greve, H., de Boeck, F., van Montagu, M., and Hernalsteens, J.P., *In vivo* random beta-glucuronidase gene fusions in *Arabidopsis thaliana*, *Proc. Natl. Acad. Sci. U.S.A.*, 88, 1, 1991.

80. Friedrich, G., and Soriano, P., Promoter traps in embryonic stem cells: a genetic screen to identify and mutate developmental genes in mice, *Genes Dev.*, 5, 1513, 1991.

81. Offringa, R., Gene targeting in plants using the *Agrobacterium* vector system, Ph.D. Thesis, Leiden University, Leiden, 1992.

82. Offringa, R., Franke-van Dijk, M.E.I., de Groot, M.J.A., van den Elzen, P.J.M., and Hooykaas, P.J.J., Nonreciprocal homologous recombination between *Agrobacterium* transferred DNA and a plant chromosomal locus, *Proc. Natl. Acad. Sci. U.S.A.*, 90, 7346, 1993.

83. Orr-Weaver, T.L., and Szostak, J.W., Yeast recombination: the association between double-strand gap repair and crossing-over, *Proc. Natl. Acad Sci. U.S.A.*, 80, 4417, 1983.

84. Thomas, K.R., Folger, K.R., and Capecchi, M.R., High frequency targeting of genes to specific sites in the mammalian genome, *Cell*, 44, 419, 1986.

85. Song, K.-Y., Schwartz, F., Maeda, N., Smithies, O., and Kucherlapati, R., Accurate modification of a chromosomal plasmid by homologous recombination in human cells, *Proc. Natl. Acad. Sci. U.S.A.*, 84, 6820, 1987.

86. Adair, G.M., Nairn, R.S., Wilson, J.H., Seidman, M.M., Brotherman, K.A., MacKinnon, C., and Scheerer, J.B., Targeted homologous recombination at the endogenous adenine phosphoribosyltransferase locus in Chinese hamster cells, *Proc. Natl. Acad. Sci. U.S.A.*, 86, 4574, 1989.

87. Gal, S., Pisan, B., Hohn, T., Grimsley, N., and Hohn, B., Agroinfection of transgenic plants leads to viable cauliflower mosaic virus by intermolecular recombination, *Virology*, 187, 525, 1992.

88. Zheng, H., Hasty, P., Brenneman, M.A., Grompe, M., Gibbs, R.A., Wilson, J.H., and Bradley, A., Fidelity of targeted recombination in human fibroblasts and murine embryonic stem cells, *Proc. Natl. Acad. Sci. U.S.A.*, 88, 8067, 1991.

89. Thomas, K.R., Deng, C., and Capecchi, M.R., High-fidelity gene targeting in embyogenic stem cells by using sequence replacement vectors, *Mol. Cell. Biol.*, 12, 2919, 1992.

90. Offringa, R., Van den Elzen, P.J.M., and Hooykaas, P.J.J., Gene targeting in plants using the *Agrobacterium* vector system, *Transgenic Res.*, 1, 114, 1992.

91. Paszkowski, J., Peterhans, A., Bilang, R., and Filipowicz, W., Expression in transgenic tobacco of the bacterial neomycin phosphotranferase gene modified by intron insertions of various sizes, *Plant Mol. Biol.*, 19, 825, 1992.

92. Hrouda, M., and Paszkowski, J., personal communications, 1993. High fidelity extrachromosomal recombination and gene targeting in plants, *Mol. Gen. Genet.*, 243, 106, 1994.

93. Risseeuw, E., personal communications, 1993.

94. Moffatt, B., and Somerville, C., Positive selection for male-sterile mutants of *Arabidopsis* lacking adenine phosphoribosyl transferase activity, *Plant Physiol.*, 86, 1150, 1988.

95. Gabard, J., Marion-Poll, A., Cherel, I., Meyer, C., Muller, A.J., and Caboche, M., Isolation and characterization of *Nicotiana plumbagenifolia* nitrate reductase deficient mutants: genetic and biochemical analysis of the nia complementation group, *Mol. Gen. Genet.*, 209, 596, 1987.

96. Rose, A.B., Casselman, A.L., and Last, R.L., A phosphoribosylanthranilate transferase gene is defective in blue fluorescent *Arabidopsis thaliana* tryptophan mutants, *Plant Physiol.*, 100, 582, 1992.

97. Bartolomei M.S., and Corden, J.L., Localization of an alpha-amanitin resistance mutation in the gene encoding the largest subunit of mouse RNA polymerase II, *Mol. Cell. Biol.*, 7, 586, 1987.

98. Steeg, C.M., Ellis, J., and Rernstein, A., Introduction of specific point mutations into RNA polymerase II by gene targeting in mouse embryonic stem cells: evidence for a DNA mismatch repair mechanism, *Proc. Natl. Acad. Sci. U.S.A.*, 87, 4680, 1990.

99. Ohl, S., unpublished data, 1992.

100. Joyner, A. L., Skarnes, W.C., and Rossant, J., Production of a mutation in mouse En-2 gene by homologous recombination in embryonic stem cells, *Nature*, 338, 153, 1989.

101. Zimmer, A., and Gruss, P., Production of chimaeric mice containing embryonic stem (ES) cells carrying a homeobox Hox 1.1 allele mutated by homologous recombination, *Nature*, 338, 150, 1989.

102. Zijlstra, M., Li, E., Sajjadi, F., Subramani, S., and Jaenisch, R., Germ-line transmission of a disrupted β2-microglobulin gene produced by homologous recombination in embryonic stem cells, *Nature*, 342, 435, 1989.

103. Jasin, M., and Berg, P., Homologous integration in mammalian cells without target gene selection, *Genes Dev.*, 2, 1353, 1988.

104. Sedivy, J.M., and Sharp, P.A., Positive genetic selection for gene disruption in mammalian cells by homologous recombination, *Proc. Natl. Acad. Sci. U.S.A.*, 86, 227, 1989.

105. Te Riele, H.T., Maandag, E.R., Clarke, A., Hooper, M., and Berns, A., Consecutive inactivation of both alleles of the pim-l proto-oncogene by homologous recombination in embryonic stem cells, *Nature*, 348, 649, 1990.

106. De Groot, M.J.A., Studies on homologous recombination in *Nicotiana tabacum*, Ph.D. Thesis, Leiden University, Leiden, 1992.

107. Mansour, S.L., Thomas, K.R., and Capecchi, M.R., Disruption of the proto-oncogene int-2 in mouse embryo-derived stem cells: a general strategy for targeting mutations to non-selectable genes, *Nature*, 336, 348, 1988.

108. Schröder, G., Waffenschmidt, S., Weiler, E., and Schröder, J., The T-region of Ti plasmids codes for an enzyme synthesizing indole-3-acetic acid, *Eur. J. Biochem.*, 138, 387, 1984.

109. Thomashow, L., Reeves, S., and Thomashow, M., Crown gall oncogenesis: evidence that a T-DNA gene from the *Agrobacterium* Ti plasmid pTiA6 encodes an enzyme that catalyses synthesis of indoleacetic acid, *Proc. Natl. Acad. Sci. U.S.A.*, 81, 5071, 1984.

110. Depicker, A.G., Jacobs, A.M., and Van Montagu, M., A negative selection scheme for tobacco protoplast-derived cells expressing the T-DNA gene, *Plant Cell Rep.*, 7, 63, 1988.

111. Karlin-Neumann, G.A., Brusslan, J.A., and Tobin, E.M., Phytochrome control of the tms2 gene in transgenic *Arabidopsis* - a strategy for selecting mutants in the signal transduction pathway, *Plant Cell*, 3, 573, 1991.

112. Yagi, T., Ikawa, Y., Yoshida, K., Shigetani, Y., Takeda, N., Mabuchi, I., Yamamoto, T., and Aizawa, S., Homologous recombination at c-fyn locus of mouse embryonic stem cells with use of diphtheria toxin A-fragment gene in negative selection, *Proc. Natl. Acad. Sci. U.S.A.*, 87, 9918, 1990.

113. Koltunow, A.M., Truettner, J., Cox, K.H., Wallroth, M., and Goldberg, R.B., Different temporal and spatial gene expression patterns occur during anther development, *Plant Cell*, 2, 1201, 1990.

114. Czako, M., and An, G., Expression of DNA coding for diphtheria toxin chain A is toxic to plant cells, *Plant Physiol.*, 95, 687, 1991.

115. Yamaizumi, M., Mekada, E., Uchida, T., and Okada, Y., One molecule of diphtheria toxin fragment A introduced into a cell can kill the cell, *Cell*, 15, 245, 1978.

116. Ward, E.R., Ryals, J.A., and Miflin, B.J., Chemical regulation of transgene expression in plants, *Plant Mol. Biol.*, 22, 361, 1993.

117. Maxwell, F., Maxwell, I.H., and Glode, M.L., Cloning, sequence determination, and expression in transfected cells of the coding sequence for the tox 176 attenuated diphtheria toxin A chain, *Mol. Cell. Biol.*, 7, 1576, 1987.

118. Breitman, M.L., Rombola, H., Maxwell, I.H., Klintworth, G.K., and Bernstein, A., Genetic ablation in transgenic mice with an attenuated diphtheria toxin A gene, *Mol. Cell. Biol.*, 10, 474, 1990.

119. Guerra, D., Xiang, C., Beremand, P., and Marton, L., The development of an acyl carrier protein homologous recombination/gene replacement in *Arabidopsis*, *J. Cell. Biochem.*, suppl., 71, 1991.

120. Xiang, C., and Guerra, D.J., The anti-nptII gene, a potential negative selectable marker for plants, *Plant Physiol.*, 102, 287, 1993.

121. Marton, L., personal communications, 1993. Czabo, M., and Marton, L., The herpes simplex virus thymidine kinase gene as a conditional negative-selection marker gene in *Arabidopsis thaliana, Plant Physiol.,* 104, 1067, 1994.

122. Nussaume, L., Vincentz, M., and Caboche, M., Constitutive nitrate reductase — a dominant conditional marker for plant genetics, *Plant J.*, 1, 267, 1991.

123. Danielsen, S., Kilstrup, M., Barilla, K., Jochimsen, B., and Neuhard, J., Characterization of the *Escherichia coli* codBA operon encoding cytosine permease and cytosine deaminase, *Mol. Microbiol.*, 6, 1335, 1992.

124. Bendich, A., Getler, H., and Brown, G., A synthesis of isotopic cytosine and a study of its methabolism in rats, *J. Biol. Chem.*, 177, 565, 1949.

125. Ross, C., Comparison of incorporation and metabolism of RNA pyrimidine nucleotide precursors in leaf tissues, *Plant Physiol.*, 40, 65, 1965.

126. Mullen, C.A., Kilstrup, M., and Blaese, R.M., Transfer of the bacterial gene for cytosine deaminase to mammalian cells confers lethal sensitivity to 5-fluorocytosine: a negative selection system, *Proc. Natl. Acad. Sci. U.S.A.*, 89, 33, 1992.

127. Stougaard, J., Substrate-dependent negative selection in plants using a bacterial cytosine deaminase gene, *Plant J.*, 3, 755, 1993.

128. Does, M.P., Dekker, B.M.M., De Groot, M.J.A., and Offringa, R., A quick method to estimate the T-DNA copy number in transgenic plants at an early stage after transformaion, using inverse PCR, *Plant Mol. Biol.*, 17, 151, 1991.

129. Feldmann, K.A., T-DNA insertion mutagenesis in *Arabidopsis*: mutational spectrum, *Plant J.*, 1, 71, 1991.

130. Lindsey, K., Wei, W., Clarke, M.C., McArdle, H.F., Rooke, L.M., and Topping, J.F., Tagging genomic sequences that direct expression by activation of a promoter trap in plants, *Transgenic Res.*, 2, 33, 1993.

131. Van Haaren, M.J.J., Sedee, N.J.A., Krul, M., Schilperoort, R.A., and Hooykaas, P.J.J., Function of heterologous and pseudo border repeats in T region transfer via the octopine virulence system of *Agrobacterium tumefaciens*, *Plant Mol. Biol.*, 11, 773, 1988.

132. Hasty, P., Rivera-Pérez, J., and Bradley, A., The role and fate of DNA ends for homologous recombination in embryonic stem cells, *Mol. Cell. Biol.*, 12, 464, 1992.

133. Kumar, S., and Simons, J.P., The effects of terminal heterologies on gene targeting by insertion vectors in embryonic stem cells, *Nucleic Acids Res.*, 21, 1541, 1993.

134. Negrutiu, I., Shillito, R., Potrykus, I., Biasini, G., and Sala, F., Hybrid genes in the analysis of transformation conditions, *Plant Mol. Biol.*, 8, 363, 1987.

135. Damm, B., Schmidt, R., and Willmitzer, L., Efficient transformation of *Arabidopsis thaliana* via direct DNA transfer to protoplasts, *Mol. Gen. Genet.*, 217, 6, 1989.

136. Krens, F.A., Mans, R.M.W., Van Slogteren, T.M.S., Hoge, J.H.C., Wullems, G.J., and Schilperoort, R.A., Structure and expression of DNA transferred to tobacco via transformation of protoplasts with Ti-plasmid DNA: co-transfer of T-DNA and non-T-DNA sequences, *Plant Mol. Biol.*, 5, 223, 1985.

137. Potrykus, I., Paszkowski, J., Saul, M., Petruska, I., and Shillito, D., Molecular and general genetics of a hybrid foreign gene introduced into tobacco by direct gene transfer, *Mol. Gen. Genet.*, 199, 169, 1985.

138. Simon, J.R., and Moore, P.D., Homologous recombination between single-stranded DNA and chromosomal genes in *Saccharomyces cerevisiae*, *Mol. Cell. Biol.*, 7, 2329, 1987.

139. Van Deursen, J., Lovellbadge, R., Oerlemans, F., Schepens, J., and Wieringa, B., Modulation of gene activity by consecutive gene targeting of one creatine kinase M-allele in mouse embryonic stem cells, *Nucleic Acids Res.*, 19, 2637, 1991.

140. Leutwiler, L.S., Hough, B.R., and Meyerowitz, E.M., The DNA of *Arabidopsis thaliana*, *Mol. Gen. Genet.*, 194, 15, 1984.

141. Meyerowitz, E.M., and Pruit, R.E., *Arabidopsis thaliana* and plant molecular genetics, *Science*, 229, 1214, 1985.

142. Zheng, H., and Wilson, J.H., Gene targeting in normal and amplified cell lines, *Nature*, 344, 170, 1990.

143. Lin, F.-L., Sperle, K., and Sternberg, N., Recombination in mouse L cells between DNA introduced into cells and homologous chromosomal sequences, *Proc. Natl. Acad. Sci. U.S.A.*, 82, 1391, 1985.

144. Capecchi, M.R., Altering the genome by homologous recombination, *Science*, 244, 1288, 1989.

145. Hasty, P., Rivera-Pérez, J., and Bradley, A., The length of homology required for gene targeting in embryonic stem cells, *Mol. Cell. Biol.*, 11, 5586, 1991.

146. Deng, C., and Capecchi, M.R., Reexamination of gene targeting frequency as a function of the extent of homology between the targeting vector and the target locus, *Mol. Cell. Biol.*, 12, 3365, 1992.

147. Miller, C.C., McPheat, J.C., and Potts, W.J., Targeted integration of the Ren-lD locus in mouse embryonic stem cells, *Proc. Natl. Acad. Sci. U.S.A.,* 89, 5020, 1992.

148. Te Riele, H., Maandag, E.R., and Berns, A., Highly efficient gene targeting in embryonic stem cells through homologous recombination with isogenic DNA constructs, *Proc. Natl. Acad. Sci. U.S.A.*, 89, 5128, 1992.

149. Scott, R.J., and Draper, J., Transformation of carrot tissues derived from proembryogenic suspension cells: a useful model system for gene expression studies in plants, *Plant Mol. Biol.*, 8, 265, 1987.

150. Vasil, V., Castillo, A.M., Fromm, M.E., and Vasil, I.K., Herbicide-resitant fertile transgenic wheat plants obtained by microprojectile bombardement of regenerable embryogenic callus, *Bio/technol.*, 10, 667, 1992.

151. Christou, P., Ford, T.L., and Kofron, M., Production of transgenic rice (Oryza sativa L) plants from agronomically important Indica and Japonica varieties via electric discharge particle acceleration of exogenous DNA into immature zygotic embryos, *Bio/technol.,* 9, 957, 1991.

152. Schoeder, H.E., Schotz, A.H., Wardeley-Richardson, T., Spencer, D., and Higgins, T.J.V., Transformation and regeneration of two cultivars of pea (Pisum sativum L.), *Plant Physiol.*, 101, 751, 1993.

153. Buerstedde, J.M., and Takeda, S., Increased ratio of targeted to random integration after transfection of chicken B-cell lines, *Cell*, 67, 179, 1991.

154. Das, O.P., Levi-Minzi, S., Koury, M., Benner, M., and Messing, M., A somatic gene rearrangement contributing to genetic diversity in maize, *Proc. Natl. Acad. Sci. U.S.A.*, 87, 7809, 1990.

155. Köhler, F., Cardon, G., Pöhlman, M., Gill, R., and Scheider, O., Enhancement of transformation rates in higher plants by low-dose irradiation: are DNA repair systems involved in the incorporation of exogenous DNA into the plant genome?, *Plant Mol. Biol.*, 12, 189, 1989.

156. Benediktsson, I., Köhler, F., and Scheider, O., Transient and stable expression of marker genes in cotransformed *Petunia* protoplasts in relation to X-ray and UV-irradiation, *Transgenic Res.*, 1, 38, 1991.

157. Cerovic, G., Bozin, D., and Dimitrijevic, B., Mismatch-specific DNA breakdown in nuclear extract from tobacco (*Nicotiana tabacum*) callus, *Plant Mol. Biol.*, 17, 887, 1991.

158. Inamdar, N.M., Zhang, Z.-Y., Brough, C.L., Gardiner, W.E., Bisaro, D.M., and Ehrlich, M., Transfection of heteroduplexes containing uracil.guanine or thymine.guanine mispairs into plant cells, *Plant Mol. Biol.*, 20, 123, 1992.

159. Murphy, T.M., Martin, C.P., and Kami, J., Endonuclease activity from tobacco nuclei specific for ultraviolet radiation-damaged DNA, *Physiol. Plant.*, 87, 417, 1993.

160. Heath-Pagliuso, S., Cole, A.D., and Kmiec, E.B., Purification and characterization of a type-I topoisomerase from cultured tobacco cells, *Plant Physiol.*, 94, 599, 1990.

161. Kieber, J.J., Lopez, M.F., Tissier, A.F., and Signer, E., Purification and properties of DNA topoisomerase I from broccoli, *Plant Mol. Biol.*, 18, 865, 1992.

162. Kieber, J.J., Tissier, A.F., and Signer, E.R., Cloning and characterization of an *Arabidopsis thaliana* topoisomerase I gene, *Plant Physiol.*, 99, 1493, 1992.

163. Cerutti, H., Osman, M., Grandoni, P., and Jagendorf, A.T., A homolog of *Escherichia coli* RecA protein in plastids of higher plants, *Proc. Natl. Acad. Sci. U.S.A.,* 89, 8068, 1992.

164. Pang, Q., Hays, J.B., and Rajagopal, I., A plant cDNA that partially complements *Escherichia coli* recA mutations predicts a polypeptide not strongly homologous to RecA protein, *Proc. Natl. Acad. Sci. U.S.A.,* 89, 8073, 1992.

165. Pang, Q., Hays, J.B., and Rajagopal, I., Two cDNAs from the plant *Arabidopsis thaliana* that partially restore recombination proficiency and DNA-damage resistance to *E. coli* mutants lacking recombination-intermediate-resolution activities, *Nucleic Acids Res.,* 21, 1647, 1993.

166. Hayes, R.J., Petty, I.T.D., Coutts, R.H.A., and Buck, K.W., Gene amplification and expression in plants by a replicating geminivirus vector, *Nature*, 334, 179, 1988.

167. Ward, A., Etessami, P., and Stanley, J., Expression of a bacterial gene in plants mediated by infectious geminivirus DNA, *EMBO J.*, 7, 1583, 1988.

168. Timmermans, M.C.P., Das, O.P., and Messing, J., Trans replication and high copy numbers of wheat dwarf virus in maize cells, *Nucleic Acids Res.*, 20, 4047, 1992.

169. Orr-Weaver, T.L., Szostak, J.W., and Rothstein, R.J., Yeast transformation: a model system for the study of recombination, *Proc. Natl. Acad. Sci. U.S.A.*, 78, 6354, 1981.

170. Jasin, M., de Villiers, J., Weber, F., and Schaffner, W., High frequency of homologous recombination in mammalian cells between endogenous and introduced SV40 genomes, *Cell*, 43, 695, 1985.

171. Engels, W.R., Johnson-Schiltz, D.M., Eggleston, W.B., and Sved, J., High-frequency P element loss in *Drosophila* is homolog dependent, *Cell*, 62, 515, 1990.

172. Gloor, G.B., Nassif, N.A., Johnson-Schlitz, D.M., Preston, C.R., and Engels, W.R., Targeted gene replacement in *Drosophila* via P element-induced gap repair, *Science*, 253, 1110, 1991.

173. Plasterk, R.H.A., and Groenen, J.T.M., Targeted alterations of the *Caenorhabditis elegans* genome by transgene instructed DNA double strand break repair following Tcl excision, *EMBO J.*, 11, 287, 1992.

Chapter 5

Gene Targeting in Mammalian Development and Physiology

Thomas Lufkin

Brookdale Center for Molecular Biology

The Mount Sinai Medical Center

New York, NY

Contents

0-8493-8950-X/95/$0.00+$.50

I. Introduction

The isolation of genes with a role in embryonic development has increased dramatically in recent years, owing partially to the amenable genetics of *Drosophila melanogaster* and *Caenorhabditis elegans,*[1] and to technical advances in molecular biology (e.g., PCR). Certain of these genes appear to have been conserved during evolution, and homologous genes have been identified in higher organisms such as mice and humans.[2-6] Although the genetic function has often been determined in *D. melanogaster* or *C. elegans*, the role of the mammalian homologue has usually remained unclear, owing principally to a lack of available genetic mutants. Until recently, the genetic manipulation of the mouse was limited to random chemical or radiation–induced mutagenesis and standard transgenic technology, both impractical for the generation of specific mutations.[7] However, recent advances in the area of homologous recombination in tissue culture cells and the isolation of pluripotent embryonic stem cells has circumvented this limitation and greatly advanced murine developmental genetics.[7-15] The analysis of mice carrying targeted mutations in specific genes has greatly enhanced our understanding of the role certain genes play in development. A largely unanticipated outcome, however, of the numerous murine genetic mutations recently generated is the apparent high degree of genetic redundancy between members of gene families.[16,17] Extensive functional redundancy might have been predicted based upon work in other organisms, as well as in population genetics.[17-22] Future efforts, in many cases, will likely focus on the analysis of mice carrying combinations of mutated gene family members.

This chapter begins with some of the key technical breakthroughs of murine gene targeting. I have not attempted to summarize all the recently generated mouse mutants that affect normal development. Instead, I have focused on a few different classes of genes (transcriptional regulators, growth factors, and a metabolic enzyme), which should serve as models for the use of gene targeting in the study of mammalian development and physiology.

II. The study of mammalian development enters a new era

A. Establishment of pluripotent stem cell lines

The isolation of pluripotent murine embryonic stem (ES) cell lines was first described over ten years ago.[23-25] This was initially accomplished through the isolation and cultivation of the inner cell mass (ICM) on a layer of supporting fibroblast feeder cells. The ICM is the group of embryonic cells that will give rise to, in part, the embryo proper at later stages of embryogenesis.[26] In one case, the ICM was immunosurgically removed from the blastocyst (embryonic day 3.5 embryo, E3.5) prior to *in vitro* culturing;[24] in the other case, blastocysts were delayed in their normal uterine implantation by ovariectomy of the mother, and the ICM was subsequently dissected away from these later stage embryos while in tissue culture.[23] Since these initial landmark reports, a number of ES cell lines has been isolated.[25] A critical characteristic of the ES cell as a tool in gene transfer studies is its ability to colonize the genital ridge of the embryo and contribute viable cells to the germline of the ensuing animal. It is this feature that permits the introduction of modified genetic information into the murine gene pool. The most common method for the introduction of ES cells into recipient embryos is injection of ES cells into the blastocoel cavity, where they subsequently integrate into the ICM to generate a chimeric embryo. Alternate approaches based upon aggregation of ES cells with morula stage embryos,[27,28] as well as the aggregation of ES cells with compromised tetraploid embryos,[29] have been successful. In the latter case, however, the offspring did not survive long after birth.

B. Transgenesis with ES cells

ES cell-mediated germ line transmission of an exogenous transgene was first performed following introduction of exogenous DNA into ES cells via calcium phosphate transfection[30] or by retroviral infection.[31] While these transgenes were not directed to any particular genetic locus, they demonstrated the feasibility of using ES cells as vectors for the introduction of foreign DNA into the mouse gene pool. Germline transmission of a genetic alteration mediated by homologous recombination was initially performed on the hypoxanthine phosphotransferase (HPRT) gene.[32,33] The factors that affect successful targeting of a gene and subsequent germline transmission of the modified ES cell are described elsewhere.[25]

C. Introduction of a LacZ marker gene to follow gene expression from the targeted locus

Genetic material (aside from the selectable marker) has been successfully introduced during homologous recombination in ES cells (Figure 1a-c).[34,35] The *Escherichia coli* β-galactosidase enzyme encoded by the LacZ gene has been inserted during homologous recombination and used as a marker to follow expression from the targeted locus. This can be particularly informative in cases where questions of cell autonomy or migration are being addressed or when an autoregulatory mechanism for the targeted locus has been predicted.[36] A simple chromogenic staining reaction[37] on either ES cells, embryoid bodies, entire small embryos, or tissue sections allows rapid *in situ* detection of β-galactosidase enzyme activity (see Figure 2 and Figure 3). In addition to its high resolution (cellular) and rapidity, the β-galactosidase assay appears to be more sensitive than other methods, such as RNA *in situ* hybridization, at estimating low levels of gene expression. However, an important caveat to keep in mind when interpreting β-galactosidase staining patterns is that enzyme activity is being assayed (not transcription or transcript abundance), and very little is known about the half-life of the β-galactosidase protein or mRNA. In addition, the introduction of the LacZ-neo sequences could contribute cryptic DNA regulatory sequences or perturb the spacing of endogenous ones, resulting in altered or ectopic domains of expression. While these are potential pitfalls, they do not appear to have posed significant problems in the cases where LacZ has been used to follow expression from the targeted locus (Figure 2 and Figure 3).[36,38]

III. Disruption of developmental control genes

A. Transcriptional regulators

1. Homeodomain-containing proteins

The *Drosophila* HOM-C complex, which is comprised of both the Antennapedia and Bithorax complexes, contains genes that have been shown to determine parasegment identity and are referred to as the homeotic genes.[5] Both "gain-of-function" and "loss-of-function" mutations have shown the homeotic genes to be able to transform the identity of one body region into the likeness of another. Murine homologues to the genes of HOM-C have been identified and are referred to as *Hox* genes, which comprise a family of 38 members separated into 4 independent clusters, each residing on a different chromosome.[39] The analysis of *Hox* gene expression patterns in developing mouse embryos suggested that they might play a role in regional patterning of the embryonic body plan (Figure 4). This is based in part on their overlapping expression domains, which differ principally at their anterior boundaries.[40]

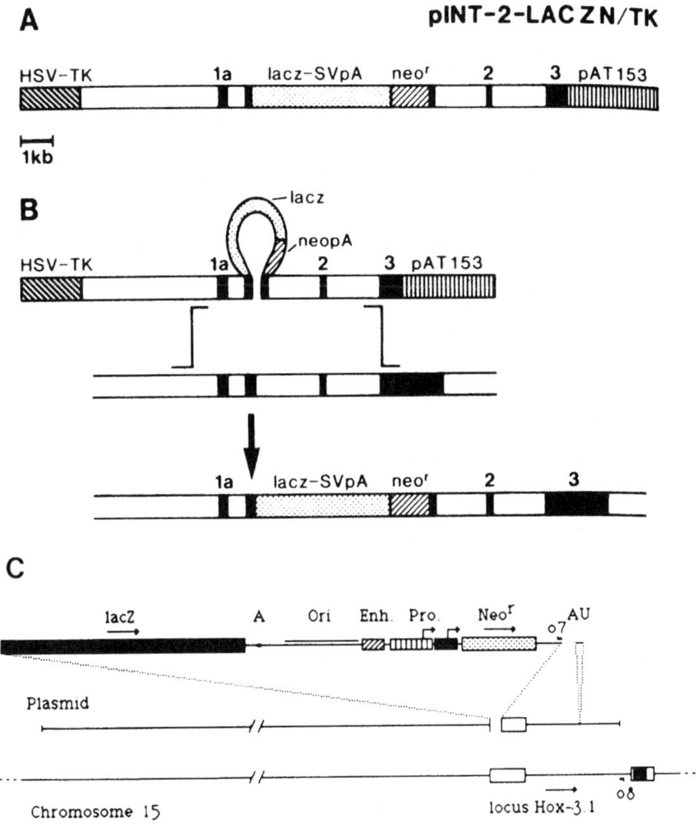

***Figure 1.** LacZ-neo targeting vectors. (**A**) The LacZ-neo targeting construct used in the disruption of int-2 (Fgf-3). Open boxes are introns and noncoding sequences. The hatched box represents the neo and HSV-TK genes, the stippled boxes represent LacZ DNA, and the vertically striped box represents plasmid sequences. (**B**) Homologous recombination between the introduced vector and the endogenous int-2 (Fgf-3) locus gives rise to the mutated allele shown below, which now expresses LacZ under the direction of int-2 (Fgf-3) transcriptional regulatory elements.[35] (reprinted with permission). (**C**) The LacZ-neo targeting construct used to disrupt the Hox-3.1 gene. The LacZ-neo containing DNA lies in a deletion within the first exon. The targeted allele contains vector and enhancer sequences in addition to the LacZ-neo sequences. The wildtype locus and the exon containing the homeobox (darkened box) are shown below. (From LeMouellic, H., Lallemand, Y., and Brulet, P.,* Proc. Natl. Acad. Sci., U.S.A., *87, 4712, 1990. With permission.)*

Earliest *Hox* gene expression is seen at the time of primitive streak formation. For certain genes, expression persists until birth and, in some tissues, even until adulthood.[40] Ectopic expression of *Hox* genes in transgenic mice ("gain-of-function" mutations) showed that these genes are able to direct regional devel-

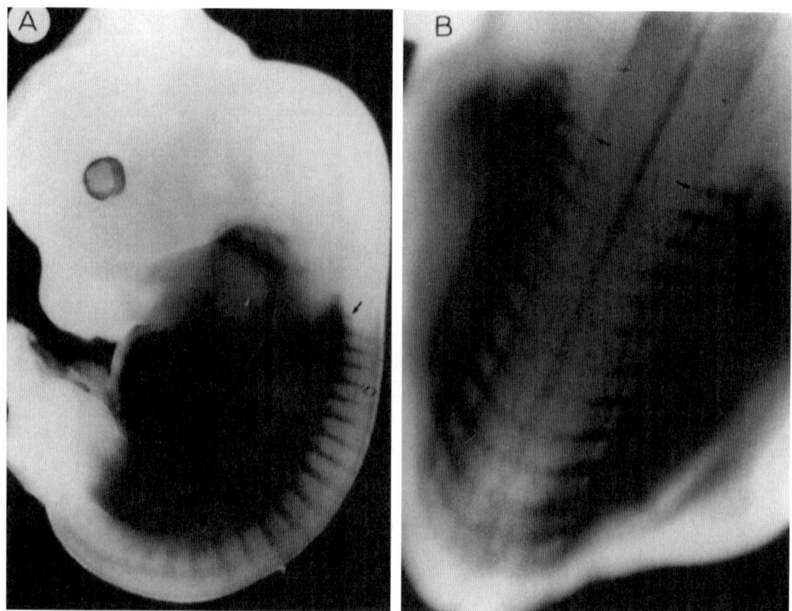

Figure 2. *LacZ expression in Hox-3.1 heterozygote embryos. (**A**) Lateral view of left side of whole-mount E12.5 labeled embryo. LacZ staining region appears dark. Arrows indicate the anterior border of LacZ expression. (**B**) Dorsal view of the same embryo in (**A**). (From LeMouellic, H.L., Lallemand, Y., and Brulet, P.,* Cell, *69, 251, 1992. With permission.)*

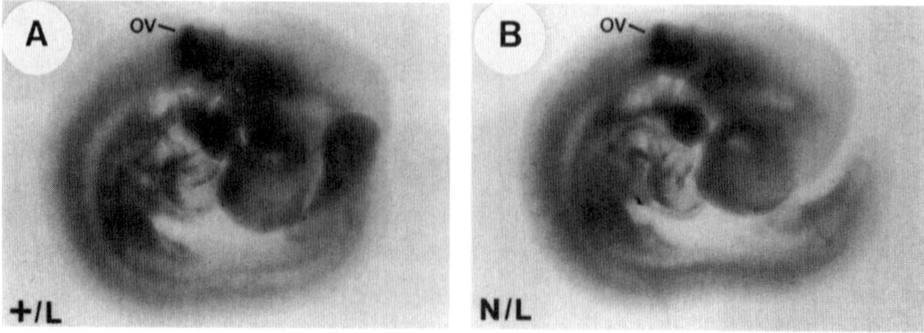

Figure 3. *LacZ expression in int-2 (Fgf-3) targeted embryos. (**A**) Right lateral view of E9.5 int-2 (Fgf-3) heterozygotes. The position of the otic vesicle (**ov**) is indicated, and the LacZ staining region appears dark. (**B**) Right lateral view of an int-2 (Fgf-3) compound heterozygote containing one neo targeted allele and one LacZ-neo targeted allele. (From Mansour, S.L., Goddard, J.M., and Capecchi, M.R.,* Development, *117, 13, 1993. With permission.)*

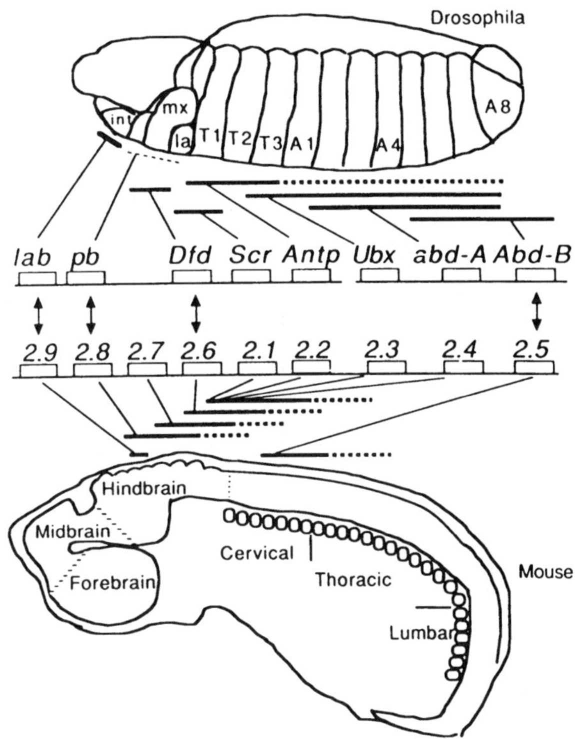

Figure 4. *Summary of HOM-C and Hox-2 expression patterns. The upper half of the figure contains a diagram of a 10 hr* Drosophila *embryo with the approximate extents of the epidermal expression domains of the HOM-C genes indicated by horizontal bars. The lower half of the figure contains a schematic diagram of a E12.5 mouse embryo, with the approximate extents of Hox-2 expression domains in the CNS indicated by the horizontal bars. (From McGinnis, W. and Krumlauf, R.,* Cell, *68, 283, 1992. With permission.)*

opment of the embryo[41-44] in a manner analogous to what has been observed with the HOM-C genes in *Drosophila*.[5]

a. *Hoxa-1 (Hox-1.6)* — The *Hoxa-1* (*Hox-1.6*) gene is located at the 3′ end of the *Hox-A* complex,[39] has one of the most extensive domains of *Hox* gene expression,[45,46] and is one of the most responsive genes to retinoic acid,[47] a putative morphagen.[48] *Hoxa-1* produces two different RNA transcripts, one of which encodes the full-length homeodomain-containing protein, and a second which encodes a C-terminal truncated protein lacking the homeodomain.[47] *Hoxa-1* expression is detected by RNA *in situ* hybridization by E7.5, where expression is seen in the primitive streak, newly formed mesoderm, and overlying neurectoderm.[45,46] *Hoxa-1* expression attains a sharp anterior border of expression by E8.0, which coincides with the preotic sulcus (presumptive rhombomere 3/4 boundary). The posterior boundary of *Hoxa-1* expression extends to the posterior end of the embryo. By E8.5, expression begins to retreat caudally in the neural tube, which is mimicked later in adjacent mesoderm and continues as a global down regulation of expression moving caudally with time. By E10.5, overall expression is markedly decreased, and by E12.5, no further *Hoxa-1* expression is detectable in the embryo. Interestingly, the two *Hoxa-1* transcripts (with and without the homeodomain) are expressed at approximately equal levels at E7.5 to E8.0, yet at all later stages, it is the homeodomain-lacking transcript that predominates.[45]

The *Hoxa-1* gene was disrupted by homologous recombination by two research groups using different targeting strategies. In one case, a DNA fragment spanning *Hoxa-1* 5′-flanking promoter sequences and the N-terminal coding region of the *Hoxa-1* protein was deleted and replaced with the neomycin-resistant (neo) gene. This mutation completely inactivated both forms of the Hoxa-1 protein (homeodomain-containing and homeodomain-lacking). [49] In the second case, the neo gene was inserted into the region encoding the homeodomain, which generated a chimeric protein composed of Hoxa-1 N-terminal sequences fused to a C-terminal sequence derived from the inserted neo gene. This mutation did not affect the Hoxa-1 homeodomain-lacking protein, which should be present in normal amounts.[50]

Mice heterozygous for either *Hoxa-1* mutation appeared normal, but *Hoxa-1* null homozygotes died soon after birth, apparently from respiratory problems. Histological analysis of *Hoxa-1* homozygotes revealed a number of defects, which were not confined to any one cell or tissue type, but were instead restricted to a discrete region of the embryo. The affected region is located at the axial level of the hindbrain rhombomeres four to six but involves many tissues and structures at this level, in addition to the hindbrain.[49,50] Structures affected in both types of *Hoxa-1* mutant mice included the occipital bones of the skull, cranial motor nerve nuclei, cranial nerve roots, ganglia, and nerve fibers, as well as the inner ear (Figure 5). Differences were observed, however, between the two types of mutations.[51,52] In the case of the complete *Hoxa-1* mutation, a delay in hindbrain neural tube closure at E9.5 was observed.[49] In the partial *Hoxa-1* mutation, defects were observed in the middle and external ears, certain cranial ganglia appeared fused, and the superior olivary complex, as well as the periodic bulges of the hindbrain rhombomeres, were absent (Figure 6).[50]

b. *Hoxa-3 (Hox-1.5)* — *Hoxa-3* (*Hox-1.5*) is a member of the *HoxA* complex and lies several kilobase pairs 5′ to *Hoxa-1*. At early stages of embryogenesis (E8.5), the *Hoxa-3* gene has an expression domain in neuroepithelium that extends from the posterior end of the embryo to an anterior boundary at the level of the presumptive rhombomere 6/7 border, which is caudal to the *Hoxa-1* (rhombomere 3/4) boundary. Expression is also detected in adjacent presomitic and somitic mesoderm, but this expression decreases with time, leaving the developing nervous system the principal domain of *Hoxa-3* expression.[53-55]

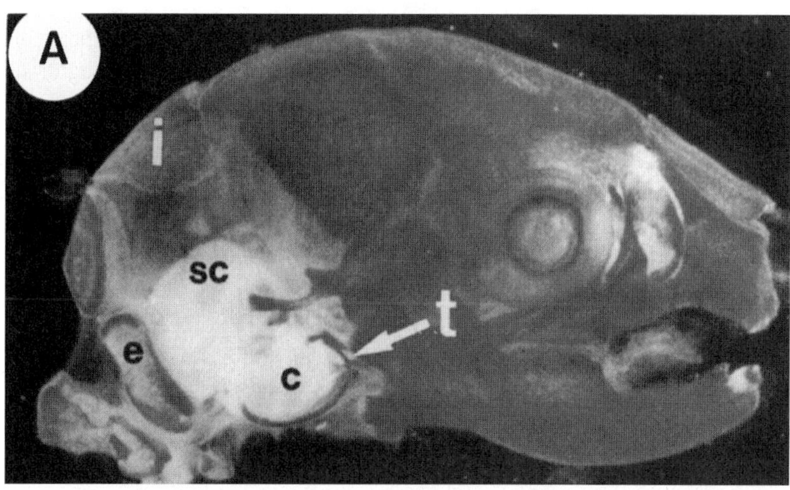

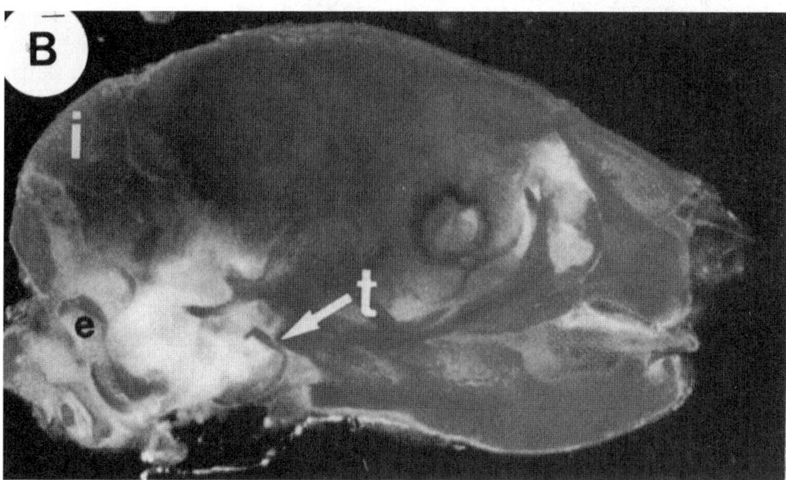

Figure 5. *Cranial region phenotype of Hoxa-1-null mutants (with both Hoxa1 proteins inactivated). (**A**) Right lateral view of a E18.5 wildtype alizarin-red stained skull. The interparietal (**i**), exoccipital (**e**), and tympanic ring (**t**) bones are indicated. The position of the semicircular canal (**sc**) and cochlea (**c**) are shown. (**B**) Right lateral view of a E18.5 Hoxa-1 mutant skull. Note the absence of the semicircular canals and cochlea and the altered shape of the exoccipital bone in the Hoxa-1 mutant. (From Lufkin, T., Dierich, A., LeMeur, M., Mark, M., and Chambon, P.,* Cell, *66, 1105, 1991. With permission).*

Later in development (E10.5 to E11.5) expression is seen in the central nervous system and in developing spinal ganglia.[53-55] By E12.5, *Hoxa-3* expression is detected in the prevertebrae, thyroid, floor of the pharynx, atrial region of the heart, lung, and kidneys.[55]

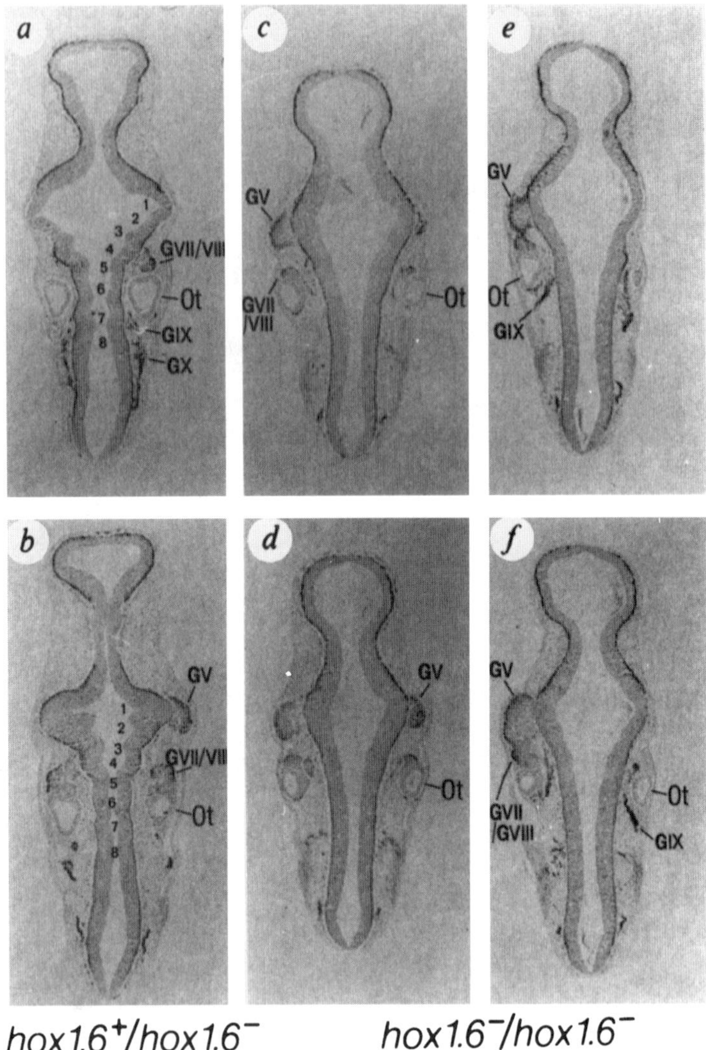

Figure 6. *Altered hindbrains of Hoxa-1-null mutants (with a single Hoxa-1 protein inactivated). (**A-F**) Coronal sections of E10.5 embryos immunoreacted with a 155K neurofilament antibody. (**A,B**) Control heterozygote embryos. (**C-F**) Mutant embryos. The sections in **A, C, E** are more dorsalrostral than the sections in **B, D, F**. Note the rhombomeric bulges numbered **1-8** in the control embryos (**A,B**) are absent in the Hoxa-1-null mutants (**C-F**). The cranial ganglia V (**GV**) and VII/VIII (**GVII/VIII**) are fused in the mutant embryos shown in **E,F**. The positions of the otic vesicle (**Ot**) and cranial ganglia V, VII-VIII, IX, and X (**GV, GVII/VIII, GIX, GX**) are indicated. (From Chisaka, O., Musci, T.S., and Capecchi, M.R.,* Nature, *355, 516, 1992. With permission.)*

The *Hoxa-3* (*Hox-1.5*) gene was disrupted by insertion of the neo gene into the homeobox.[56] This disruption generated a C-terminally truncated Hoxa-3 protein fused to sequences encoded by the inserted neo gene. Mice heterozygous for the *Hoxa-3* mutation appeared normal. Analysis of embryos in utero derived from crosses between *Hoxa-3* heterozygous parents showed approximately 25% of the embryos to be *Hoxa-3* null homozygotes, indicating that the mutation was not embryonic lethal. Newborn *Hoxa-3* null homozygotes died soon after birth, in many cases apparently from cardiopulmonary dysfunction. In certain animals, air appeared to be misrouted into the abdomen.[56]

Histological analysis of *Hoxa-3* null homozygotes at E10.5 revealed slight alterations in the appearance of the pharyngeal arches. At later stages of embryogenesis, *Hoxa-3* null homozygotes demonstrated an absence or decrease in size of the parathyroids, thymus, and thyroid. A range of cardiovascular defects were observed, including agenesis of the carotid artery and dysmorphogenesis of the heart (atrium and ventricle) compartments. Skeletal defects were observed in the *Hoxa-3* null homozygotes, which included the absence of the lesser horn of the hyoid bone, alterations in the thyroid and cricoid cartilages, as well as the maxilla, mandible, and zygomatic process of the squamosal bone. The throat musculature, larynx, trachea, and esophagus appeared abnormal and may be related to the respiratory difficulties observed immediately following birth.[56]

Interestingly, there was no observable defect in many structures that normally express *Hoxa-3*, such as the stomach, lung, kidney, spleen, central nervous system, cranial nerves, or ganglia. Whether this reflects a functional redundancy between *Hoxa-3* and other *Hox* genes expressed in these tissues remains to be determined.

Hoxa-1 and *Hoxa-3* have two of the most extensive domains of expression of the *Hox* genes, yet each gene plays a unique developmental role only in a very restricted region within this extended domain, which coincides with its rostral domain of expression. This phenomenon has been observed for other *Hox* gene disruptions[36,56,57] and will likely be a common characteristic of the *Hox* genes. The restriction of *Hox* gene function to the anterior domain of expression may be the result of the increasingly greater numbers of *Hox* genes expressed in more posterior regions of the embryo that are able to activate the same necessary target genes (functional redundancy). Alternatively, the "anterior" *Hox* genes (Hoxa-1, Hoxa-3, etc.) may be expressed in posterior regions where they play no functional role (see Section V.).

2. Retinoic acid receptors

Retinoic acid (RA) is a vitamin A derivative that can have profound effects in both deficiency and excess upon embryonic development and the maintenance of adult tissues.[58] RA has properties that are characteristic of a morphogen (a substance that directs the regional patterning of cells), and endogenous sources of RA may act to initiate or direct normal patterning of various embryonic structures such as the limb and spinal cord.[48] The effects of RA are likely mediated by two families of ligand-inducible transcriptional regulatory factors, the retinoic acid receptors (RARs) and the retinoid X receptors (RXRs), which are members of the superfamily of steroid/thyroid hormone nuclear receptors. Each receptor is expressed as multiple isoforms, owing to differential promoter usage and splicing.[59] A conserved region among the members of the nuclear receptor superfamily is the zinc-stabilized structural domain (zinc-fingers) responsible for DNA binding.[60] The RARs can be activated by binding either all-trans RA or 9-cis RA, whereas the RXRs are activated only by 9-cis RA. The activation of RA-responsive genes appears, in many cases, to be mediated by the binding of a RAR-RXR heterodimer.[59]

The RARs are encoded by three different genes, RARα, RARβ, and RARγ, and each gene has a specific pattern of expression during embryogenesis. RARα has the most widespread pattern, being expressed almost ubiquitously except for certain regions of the developing forebrain and midbrain. RARα expression is first detectable at E8.0 and continues throughout embryogenesis, persisting well into adulthood.[61] The RARα gene is expressed as two principal isoforms, RARα1 and RARα2, with RARα1 being the more ubiquitous and strongly expressed isoform.[62]

The analysis of RARα gene function was approached by generating two different kinds of mutations. In one case, the RARα gene was disrupted by insertion of the neo gene into an N-terminal region common to all RARα isoforms, which resulted in a total disruption of the RARα gene (RARα mutation).[63] In another case, the RARα1 isoform was selectively disrupted by insertion of the neo into the RARα1-specific exon (RARα1 mutation), thus leaving the other RARα isoforms (principally RARα2) unaffected.[63,64] Animals heterozygous for the RARα null mutation appeared normal and were fertile. RARα-null homozygotes displayed high perinatal lethality (60% of RARα-null homozygotes were eliminated within the first 12 to 24 hours), owing to preferential cannibalization by the dams.[63] RARα-null homozygotes that were delivered by cesarean and placed in isolation during a 24 hour period showed neither obvious physical or behavioral abnormalities nor an increased mortality relative to littermates, yet clearly, they had a characteristic recognizable by the dams. By two months of age, 90% of the RARα-null homozygotes had died or been cannibalized. Surprisingly, histological analysis of newborn RARα-null homozygotes revealed no visible abnormalities. The occasional male RARα-null homozygote, which survived longer than two months, never produced offspring, even when caged with wildtype females. Analysis of the testis of these RARα-null homozygotes showed severe degeneration of the germinal epithelium (Figure 7). The seminiferous tubules were devoid of spermatogenic cells and were filled instead with vacuolating Sertoli cells.[63]

In contrast to what was seen for total disruption of the RARα gene, mice with a mutation in the RARα1-specific isoform displayed none of the characteristics of the RARα–null homozygotes. The RARα1-null homozygotes were

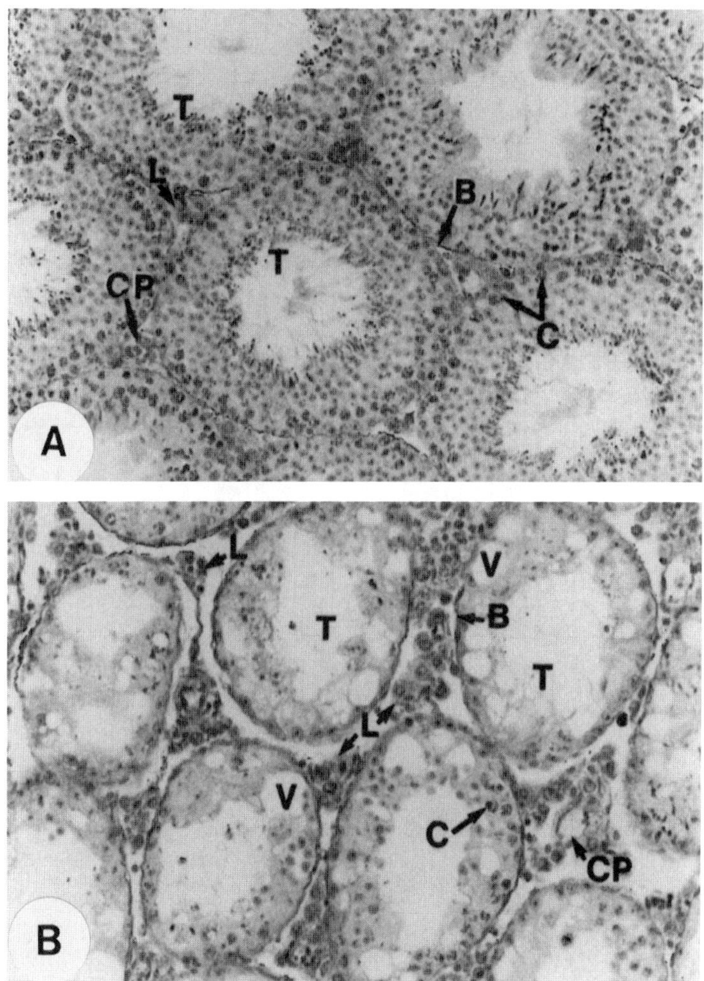

Figure 7. *Degenerative lesions in testes of RARα-null homozygotes. Histological sections through the testes of a wildtype (panel* ***A****) and an RARα-null homozygote (panel* ***B****). The parenchyma of wildtype testes is composed of seminiferous tubules (****T****) with active spermatogenesis and intertubular spaces containing capillaries (****CP****) and Leydig cells (****L****). The position of the basement membrane (****B****) is indicated. Primary spermatocytes (****C****) will eventually yield four spermatozoids. The RARα-null homozygote testes (panel* ***B****) display patchy lesions of the seminiferous tubules (****T****). The majority of the tubules lack primary spermatocytes (****C****). In addition, the seminiferous epithelium shows large, clear, rounded spaces (vacuole-like,* ***V****). In the intertubular spaces, focal hyperplasia of the Leydig cells (****L****) is observed between atrophic seminiferous tubules. Photograph kindly provided by Dr. Manuel Mark, Strasbourg.*

healthy, fertile, and displayed no visible phenotype.[63,64] This was surprising, since the RARα1 isoform is the more abundant and ubiquitously expressed of the two RARα isoforms. Analysis of the RNA levels of the other RARα isoforms and other RARs (RARβ, RARγ) in the RARα1-null homozygotes showed no detectable alterations in their expression, suggesting that the lack of a visible phenotype in the RARα1-null homozygotes was not due to compensatory expression by the other RARs.[63,64] Thus, one can conclude that there is significant functional redundancy between RARα1 and the other RARα isoforms. Furthermore, the phenotype obtained from the RARα-null mutation, while lethal, did not show any significant lesions (apart from testicular degeneration). This was unexpected, considering the extensive domain of expression of RARα, and may reflect a functional redundancy with either RARβ or RARγ. This conclusion is supported by the analysis of mice doubly homozygous for the RARα-null mutation and a null mutation in either RARβ or RARγ (Pierre Chambon and Manuel Mark, personal communications).[88]

B. Growth factors

1. *Int-2 (Fgf-3)*

The *int-2* gene (also known as *Fgf-3*) was isolated as a proto-oncogene activated following integration of mouse mammary tumor virus (MMTV) proviral DNA in virally-induced mammary tumors.[65,66] Characterization of the *int-2* gene identified it as a member of the fibroblast growth factor (FGF) family.[67,68] The *int-2* gene expresses multiple forms of mRNA, owing to alternate use of transcription initiation and polyadenylation sites; however, the protein likely encoded by these multiple forms appears identical.[69,70]

The expression of *int-2* analyzed by RNA *in situ* hybridization revealed *int-2* transcripts in multiple tissues and at different times during murine development.[71,72] At E7.5 to E8, expression was detected in the primitive streak and in cells moving from the primitive streak to extra-embryonic sites. By E9.5, *int-2* transcripts were still present in the primitive streak, but were confined to migrating cells and not to more structured mesoderm such as the somites. Expression was also detected in presumptive hindbrain neurepithelium at a level adjacent to the developing otic vesicles and in endoderm of the pharyngeal pouches.[72] At later times of development, *int-2* transcripts were detected in Purkinje cells of the cerebellum, in regions of the developing retina, in mesenchyme of developing teeth, and in sensory regions of the inner ear.[71]

The expression of *int-2* in hindbrain epithelium adjacent to the otic vesicle and, subsequently, in the otic vesicle and then developing sensory regions of the inner ear, suggests a possible role for *int-2* in development of the inner ear. Chick explants incubated with antisense *int-2* oligonucleotides or anti-int-2 antibodies failed to form a normal otic vesicle, suggesting that *int-2* was a natural inductive signal for the otic vesicle.[73]

The *int-2* gene was disrupted by introduction of a neo or neo-LacZ gene into the exon encoding the N-terminal region of the 245 amino acid int-2 protein.[35,38,70,74] Heterozygous mice appeared normal, but homozygous mutant mice showed high perinatal lethality, with less than half the expected number of homozygotes surviving the first 24 hours following birth and the majority of the remaining homozygotes dying within 2 days to 3 weeks. Autopsies, however, revealed no specific causes for the observed lethality.[38] Examination at prenatal stages showed that *int-2*-null homozygotes represented around 25% of the embryos; hence, the mutation did not appear to be embryonic lethal. *int-2*-null homozygotes were identifiable macroscopically from E12.5 onward, by a visible kinking or curling of the tail (Figure 8). The skeletons of *int-2*-null homozygote newborns had fused or abnormally shaped caudal vertebrae, which were delayed in ossification. Examination at earlier times (E11.5) showed an abnormal caudal extension of the gut tube, as well as altered positioning of somites and of the gut tube relative to the notochord.[38]

Analysis of E9.5 *int-2*-null homozygotes revealed no visible defect in the otic vesicle, nor was there any detectable alteration in the adjacent hindbrain region that expresses *int-2* (rhombomeres 5 and 6). However at later stages,

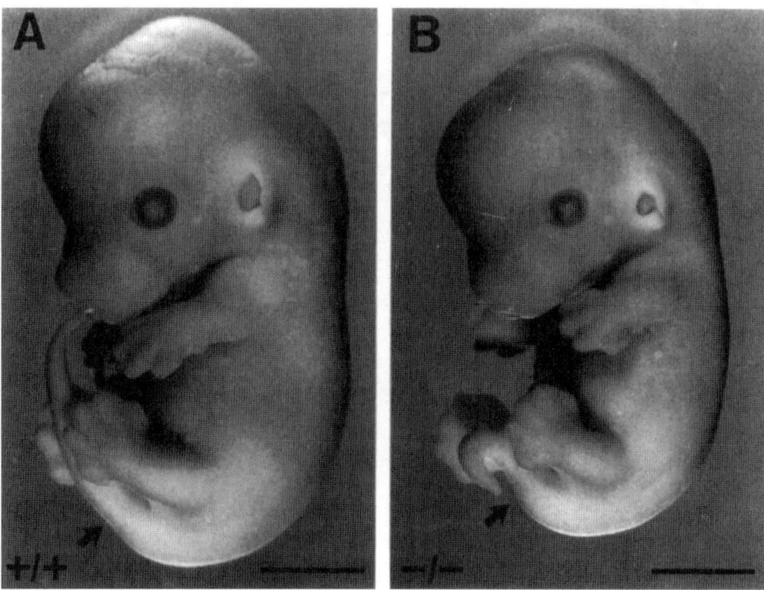

Figure 8. *Int-2 (Fgf-3) mutant embryos have abnormal tails. Left lateral view of two E13.5 littermates. (**A**) Homozygous wildtype. (**B**) Int-2 (Fgf-3)-null homozygote. Note the kink in the tail (arrow) of the int-2 (Fgf-3)-null homozygote shown in (**B**). Bar, 2mm. (From Mansour, S.L., Goddard, J.M., and Capecchi, M.R.,* Development, *117, 13, 1993. With permission.)*

visible alterations, principally a reduction in size, were observed in the endolymphatic duct and the facio-acoustic ganglia.[38] The inner ear defects were heterogeneous from one animal to another, but were characterized generally by a distention of the auditory and vestibular components (Figure 9), although differentiation of inner ear sensory epithelium appeared normal. No defects were observed in the otic capsule nor in the auditory regions of the brain.[38]

The defects in the *int-2*-null homozygote tails are consistent with the expression of *int-2* at earlier stages in the primitive streak and may reflect a role for *int-2* in either proliferation or migration of cells in the tail. The defects in the inner ears of *int-2*-null homozygotes are likely a result of failure of normal

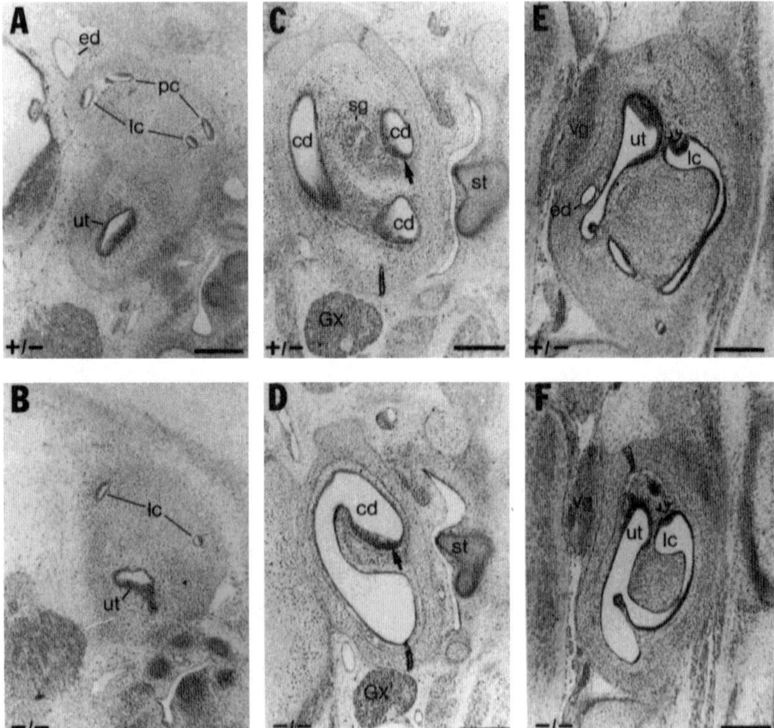

Figure 9. *Inner ear defects in E13.5 and E15.5 int-2 (Fgf-3)-null homozygote embryos. Sagittal sections through the vestibular portion of the inner ear of E13.5 heterozygous (wildtype) (**A**) and homozygous (mutant) (**B**) embryos. Bar, 300 μm. Transverse sections taken through the cochlear (**C** and **D**) and vestibular (**E** and **F**) portions of E15.5 embryos. The developing cochlear duct (in **C** and **D**) is indicated with an arrow. Bar, 200 μm. **ed**=endolympahtic duct; **ut**=utricle; **lc**=lateral canal; **pc**=posterior canal; **sg**=spiral ganglion; **cd**=cochlear duct; **st**=stapes; **GX**=tenth ganglion; **vg**=vestibular ganglion. (From Mansour, S.L., Goddard, J.M., and Capecchi, M.R.,* Development, *117, 13, 1993. With permission.)*

endolymphatic duct formation, for which normal development may be dependent upon *int-2* inductive signals emanating from the hindbrain and/or from the otic vesicle itself.[38]

Many structures that express *int-2* in wildtype animals, such as the retina, teeth, and cerebellum, were apparently unaffected in the *int-2*-null homozygotes. In addition, while the tail phenotype was 100% penetrant, only 60% of the *int-2* mutant ears appeared affected. Interestingly, within a single *int-2*-null homozygote, the inner ears were not always equally affected.[38] A weak penetrance can often be explained by variations in genetic background, yet a phenotypic difference in bilateral structures within the same animal (expressivity) points to other possibilities, such as non-genetic factors or possibly parallel signalling pathways, which in the case of *int-2*, could involve other members of the FGF family.[75]

IV. Gene disruption in physiology and metabolism

A. Glucocerebrosidase

Glucocerebrosidase (β-D-glucosyl-N-acylsphingosine glucohydrolase) is an enzyme that degrades sphingolipid glucocerebroside. An autosomally inherited deficiency in this enzyme results in the lysosomal storage disorder known as Gaucher's disease.[76]

To generate an animal model of this disease, the glucocerebrosidase gene was inactivated by the insertion of a neo cassette into the region of the gene encoding the active site of the enzyme.[77] Heterozygous mice were viable and fertile and had approximately half the normal levels of glucocerebrosidase activity compared to wildtype animals. Mice homozygous for the glucocerebrosidase-null allele had less than 4% of the wildtype glucocerebrosidase activity and died within 24 hours following birth; the lethality, however, could not be attributed to any specific lesion. The glucocerebrosidase-null homozygotes were underweight at birth and had severe respiratory problems, which is similar to what has been observed in severely affected Gaucher's patients. Analysis of the liver, spleen, bone marrow, and brain revealed macrophages with accumulated lysosomal lipids. The accumulated lipids had a similar physical appearance as the tubular lysosomal deposits seen in Gaucher's disease.[77]

V. Genetic redundancy and superfluous gene expression

A recurring observation in the increasing number of mouse mutants generated by gene targeting is the apparent extensive functional redundancy observed for certain genes, in particular for members of gene families. Although previously observed in the mouse,[18] genetic redundancy is clearly not unique to the murine system, having also been observed in yeast[78-80] (reviewed in Reference 22), the fruit fly[81] (reviewed in Reference 17), the nematode,[82] and cruciferous plants.[83]

It has been argued that *full* functional redundancy is unlikely to exist among two members of a multigene family, since this would impart no selective advantage, and either of the two genes would eventually decay into a nonfunctional or pseudogene. However, to retain a new addition to a gene family, particularly in large populations, only a very small selective advantage would have to be acquired by the newly duplicated member.[22] This could be either a subtle alteration in protein function or even a slight change in the domain of expression. In the latter case, there would clearly be a functional overlap in the subset of cells that expressed both gene family members. In such an example, targeted disruption of either of the two members individually could result in a very subtle phenotypic alteration. If the fitness reduction from the mutation were 5% (which is significant on a evolutionary timescale), one would have to examine around 20,000 mutant animals using the correct phenotypic analysis to have a high probability (95%) of clearly identifying the altered phenotype.[22] Thus, the examples of gene targeting where it has been concluded that there is "no phenotype"[64,84-86] are likely the result of either the analysis of insufficient animals or the application of the inappropriate phenotypic test. The issue of appropriate phenotypic test may pose a particularly difficult problem in analysis of the nervous system, where genes may be involved in complex adaptive functions that only have true functional relevance in the wild.

Another mechanism that could affect the severity of phenotypes obtained through gene targeting is a feed-back mechanism, which results in altered expression of other genes in the mutant animal and which might then functionally compensate for the targeted gene through their increased or altered pattern of expression. While this does not appear to be the case for most mutations where it has been examined, it has been suggested that the induction of Myf-5 may partially compensate for the targeted disruption of the MyoD gene.[85]

An alternative explanation, which is distinct from genetic redundancy, is that many genes may be expressed in regions where they have absolutely no functional role (superfluous expression), yet at the same time, have no negative impact. I will use the *Hox* genes (described above) as a possible example of this situation. The "anterior" *Hox* genes (*Hoxa-1*, *Hoxa-3*, etc.) have very extensive expression domains, but are functionally important only within a very restricted region within this domian which corresponds to their rostral boundary. Their expression in posterior regions may simply reflect the simplicity of the mechanism used to activate them at the time and in the domain (rostral) in which they are absolutely required. If there is no deleterious effect associated with "anterior" *Hox* gene expression in posterior regions, then it is possible that no regulatory mechanism was aquired to turn off their expression in regions where they play no role (nor have a negative impact!). It would be difficult to argue that the additional metabolic burden on the cell of expressing a few genes in regions where they are not absolutely required would be sufficient justification to develop a complex regulatory mechanism to turn them off. The idea that a gene is expressed only when and where it is needed may be incorrect. Superfluous or "junk" expression of a protein (expression in a region where it has no positive or negative impact) may be a common feature of genes and has already been described in a number of cases in mouse[16] and in *Drosophila* .[87] If this turns out to be a general phenomenon (as it shows signs of becoming), predictions of gene function based solely upon mRNA or protein expression patterns should be viewed with increased caution.

VI. Conclusion

Gene targeting in ES cells has transformed the field of murine developmental genetics. Much information has been obtained from the targeting of individual genes; however, functional redundancy, particularly between members of gene families, has hampered the analysis of certain developmental and signaling pathways. Furthermore, in the analysis of mutant phenotypes in the laboratory, researchers cannot hope to replicate the complexity of the natural environment; thus, many mutants will be judged asymptomatic owing to an inadequacy of suitable testing criteria or, in the case where a small fitness difference can be phenotypically tested, the generation and analysis of sufficient animals. While these potential limitations are daunting, they are clearly not prohibitive. In many instances, future work will focus upon the generation of mice carrying multiple mutations.

Acknowledgments

I thank my collegues from the Brookdale Center for Molecular Biology and Achim Gossler for critical comments on the manuscript. This is manuscript #170 from the Brookdale Center for Molecular Biology. This work was supported by the Deafness Research Foundation. TL is an Alfred P. Sloan Research Fellow and a Lucille B. Markey Foundation Scholar.

References

1. Glover, D.M. and Hames, B.D., Eds. *Genes and Embryos*, 1st ed., IRL Press, Oxford, 1989.

2. Holland, P., Ingham, P., and Krauss, S., Mice and flies head to head, *Nature*, 358, 627, 1992.

3. Holland, P., Homeobox genes in vertebrate evolution, *BioEssays*, 14, 267, 1992.

4. Kessel, M. and Gruss, P., Murine developmental control genes, *Science*, 249, 374, 1990.

5. McGinnis, W. and Krumlauf, R., Homeobox genes and axial patterning, *Cell*, 68, 283, 1992.

6. Gruss, P. and Walther, C., Pax in development, *Cell*, 69, 719, 1992.

7. Gridley, T., Insertional versus targeted mutagenesis in mice, *New Biol.*, 3, 1025, 1991.

8. Baribault, H. and Kemler, R., Embryonic stem cell culture and gene targeting in transgenic mice, *Mol. Biol. Med.*, 6, 481, 1989.

9. Capecchi, M.R., Altering the genome by homologous recombination, *Science*, 244, 1288, 1989.

10. Doi, S., Campbell, C., and Kucherlapati, R., Directed modification of genes by homologous recombination in mammalian cells, in *Transgenic Animals*, Grosveld, F. and Kollias, G., Eds., Academic Press Inc., San Diego, 1992.

11. Frohman, M.A. and Martin, G.R., Cut, paste, and save: new approaches to altering specific genes in mice, *Cell*, 56, 145, 1989.

12. Gossler, A. and Balling, R., The molecular and genetic analysis of mouse development, *Eur. J. Biochem.*, 204, 5, 1992.

13. Pascoe, W.S., Kemler, R., and Wood, S.A., Genes and functions: trapping and targeting in embryonic stem cells, *Biochim. Biophys. Acta*, 1114, 209, 1992.

14. Zimmer, A., Manipulating the genome by homologous recombination in embryonic stem cells, *Annu. Rev. Neurosci.*, 15, 114, 1992.

15. Evans, M.J., Potential for genetic manipulation of animals, *Mol. Biol. Med.*, 6, 557, 1989.

16. Erickson, H.P., Gene knockouts of c-src, transforming growth factor b1, and tenascin suggest superfluous, nonfunctional expression of proteins, *J. Cell Biol.*, 120, 1079, 1993.

17. Tautz, D., Redundancies, development, and the flow of information, *BioEssays*, 14, 263, 1992.

18. Abe, R., Foo-Phillips, M., and Hodes, R.J., Analysis of Mls-c genetics: a novel instance of genetic redundancy, *J. Exp. Med.*, 170, 1059, 1989.

19. Ohta, T., Evolution by gene duplication and compensatory advantageous mutations, *Genetics*, 120, 841, 1988.

20. Ohta, T., Role of gene duplication in evolution, *Genome*, 31, 301, 1989.

21. Ohta, T., Time for compensatory mutations under gene duplication, *Genetics*, 123, 579, 1989.

22. Brookfield, J., Can genes truly be redundant?, *Curr. Biol.,* 2, 553, 1992.

23. Evans, M.J. and Kaufman, M.H., Establishment in culture of pluripotential cells from mouse embryos, *Nature*, 292, 154, 1981.

24. Martin, G.R., Isolation of a pluripotent cell line from early mouse embryos cultured in medium conditioned by teratocarcinoma stem cells, *Proc. Natl. Acad. Sci. U.S.A.*, 78, 7634, 1981.

25. Hooper, M.L., Embryonal stem cells: introducing planned changes into the animal germline, in *Modern Genetics*, Vol. 1, 1st ed., Evans, H.J., Ed., Chur: Harwood Academic Publishers, 1992.

26. Gossler, A., Early mouse development, in *Early Embryonic Development of Animals*, Hennig, W., Ed., Springer-Verlag, Berlin, 1992.

27. Bradley, A., Production and analysis of chimaeric mice, in *Teratocarcinomas and Embryonic Stem Cells: A Practical Approach,* Robertson, E.J., Ed., IRL Press Limited, Oxford, 1987.

28. Wood, S.A., Pascoe, W.S., Schmidt, C., Kemler, R., Evans, M.J., and Allen, N.D., Simple and efficient production of embryonic stem cell-embryo chimeras by coculture, *Proc. Natl. Acad. Sci. U.S.A.*, 90, 4582, 1993.

29. Nagy, A., Gocza, E., Diaz, E.M., Prideaux, V.R., Ivanyi, E., Markkula, M., and Rossant, J., Embryonic stem cells alone are able to support fetal development in the mouse, *Development,* 110, 815, 1990.

30. Gossler, A., Doetschman, T., Korn, R., Serfling, E., and Kemler, R., Transgenesis by means of blastocyst-derived embryonic stem cells lines, *Proc. Natl. Acad. Sci. U.S.A.*, 83, 9065, 1986.

31. Robertson, E., Bradley, A., Kuehn, M., and Evans, M.J., Germ-line transmission of genes introduced into cultured pluripotential cells by retroviral vector, *Nature*, 323, 445, 1986.

32. Koller, B.H., Hagemann, L.J., Doetschman, T., Hagman, J.R., Huang, S., Williams, P.J., First, N.L., Maeda, N., and Smithies, O., Germ-line transmission of a planned alteration made in a hypoxanthine phosphotransferase gene by homologous recombination in embryonic stem cells, *Proc. Natl. Acad. Sci. U.S.A.*, 86, 8927, 1989.

33. Thompson, S., Clarke, A.R., Pow, A.M., Hooper, M.L., and Melton, D.W., Germ line transmission and expression of a corrected HPRT gene produced by gene targeting in embryonic stem cells, *Cell*, 56, 313, 1989.

34. LeMouellic, H., Lallemand, Y., and Brulet, P., Targeted replacement of the homeobox gene Hox-3.1 by the *Escherichia coli* lacZ in mouse chimeric embryos, *Proc. Natl. Acad. Sci. U.S.A.*, 87, 4712, 1990.

35. Mansour, S.L., Thomas, K.R., Deng, C., and Capecchi, M.R., Introduction of a lacZ reporter gene into the mouse int-2 locus by homologous recombination, *Proc. Natl. Acad. Sci. U.S.A.,* 87, 7688, 1990.

36. LeMouellic, H.L., Lallemand, Y., and Brulet, P., Homeosis in the mouse induced by a null mutation in the Hox-3.1 gene, *Cell*, 69, 251, 1992.

37. Sanes, J.R., Rubenstein, J.L.R., and Nicolas, J.F., Use of a recombinant retrovirus to study post-implantation cell lineage in mouse embryos, *EMBO J.*, 5, 3133, 1986.

38. Mansour, S.L., Goddard, J.M., and Capecchi, M.R., Mice homozygous for a targeted disruption of the proto-oncogene int-2 have developmental defects in the tail and inner ear, *Development*, 117, 13, 1993.

39. Scott, M.P., Vertebrate homeobox gene nomenclature, *Cell*, 71, 551, 1992.

40. Holland, P.W.H. and Hogan, B.L.M., Expression of homeobox genes during mouse development: a review, *Genes Dev.,* 2, 773, 1988.

41. Kessel, M., Balling, R., and Gruss, P., Variations of cervical vertebrae after expression of a Hox-1.1 transgene in mice, *Cell*, 61, 301, 1990.

42. Lufkin, T., Mark, M., Hart, C.P., LeMeur, M., and Chambon, P., Homeotic transformation of the occipital bones of the skull by ectopic expression of a homeobox gene, *Nature*, 359, 835, 1992.

43. Pollock, R.A., Jay, G., and Bieberich, C.J., Altering the boundaries of Hox3.1 expression: evidence for antipodal gene regulation, *Cell*, 71, 911, 1992.

44. Jegalian, B.G. and Robertis, E.M.D., Homeotic transformation in the mouse induced by overexpression of a human Hox3.3 transgene, *Cell*, 71, 901, 1992.

45. Murphy, P. and Hill, R.E., Expression of the mouse labial-like homeobox-containing genes, Hox 2.9 and Hox 1.6, during segmentation of the hindbrain, *Development,* 111, 61, 1991.

46. Sundin, O.H., Busse, H.G., Rogers, M.B., Gudas, L.J., and Eichele, G., Region-specific expression in early chick and mouse embryos of Ghox-lab and Hox 1.6, vertebrate homeobox-containing genes related to *Drosophila* labial, *Development*, 108, 47, 1990.

47. LaRosa, G.J. and Gudas, L.J., Early retinoic acid-induced F9 teratocarcinoma stem cell gene ERA-l: alternate splicing creates transcripts for a homeobox-containing protein and one lacking the homeobox, *Mol. Cell. Biol.*, 8, 3906, 1988.

48. Maden, M. and Tickle, C., Retinoic acid and vertebrate development, *Semin. Dev. Biol.*, 2, 151, 1991.

49. Lufkin, T., Dierich, A., LeMeur, M., Mark, M., and Chambon, P., Disruption of the Hox-1.6 homeobox gene results in defects in a region corresponding to its rostral domain of expression, *Cell,* 66, 1105, 1991.

50. Chisaka, O., Musci, T.S., and Capecchi, M.R., Developmental defects of the ear, cranial nerves and hindbrain resulting from targeted disruption of the mouse homeobox gene Hox-1.6, *Nature*, 355, 516, 1992.

51. Noden, D.M. and Van de Water, T.R., Genetic analysis of mammalian ear development, *Trends Neurosci.,* 15, 235, 1992.

52. Hogan, B. and Wright, C., The making of the ear, *Nature*, 355, 494, 1992.

53. Gaunt, S.J., Homoeobox gene Hox-1.5 expression in mouse embryos: earliest detection by *in situ* hybridization is during gastrulation, *Development*, 101, 51, 1987.

54. Fainsod, A., Awgulewitsch, A., and Ruddle, F.H., Expression of the murine homeobox gene Hox 1.5 during embryogenesis, *Dev. Biol.*, 124, 125, 1987.

55. Gaunt, S.J., Mouse homeobox gene transcripts occupy different but overlapping domains in embryonic germ layers and organs: a comparison of Hox-3.1 and Hox-1.5, *Development*, 103, 135, 1988.

56. Chisaka, O. and Capecchi, M.R., Regionally restricted developmental defects resulting from targeted disruption of the mouse homeobox gene hox1.5, *Nature*, 350, 473, 1991.

57. Ramirez-Solis, R., Zheng, H., Whiting, J., Krumlauf, R., and Bradley, A., Hoxb-4 (Hox-2.6) mutant mice show homeotic transformation of a cervical vertebra and defects in the closure of the sternal rudiments, *Cell*, 73, 279, 1993.

58. Morriss-Kay, G.M., Ed., *Retinoids in Normal Development and Teratogenesis,* 1st ed, Oxford University Press, Oxford, 1992.

59. Leid, M., Kastner, P., and Chambon, P., Multiplicity generates diversity in the retinoic acid signalling pathways, *Trends Biochem. Sci.,* 17, 427, 1992.

60. Green, S. and Chambon, P., Nuclear receptors enhance our understanding of transcriptional regulation, *Trends Genet.,* 4, 309, 1988.

61. Ruberte, E., Dolle, P., Chambon, P., and Morriss-Kay, G., Retinoic acid receptors and cellular retinoid binding proteins II. Their differential pattern of transcription during early morphogenesis in mouse, *Development*, 111, 45, 1991.

62. Leroy, P., Krust, A., Zelent, A., Mendelsohn, C., Garnier, J.M., Kastner, P., Dierich, A., and Chambon, P., Multiple forms of the mouse retinoic acid receptor α are generated by alternate splicing and differential induction by retinoic acid, *EMBO J.*, 10, 59, 1991.

63. Lufkin, T., Lohnes, D., Mark, M., Dierich, A., Gorry, P., Gaub, M.P., LeMeur, M.P., and Chambon, P., High postnatal lethality and testis degeneration in retinoic acid receptor α (RARα) null mice, *Proc. Natl. Acad. Sci. U.S.A.*, 90, 7225, 1993.

64. Li, E., Sucov, H.M., Lee, K., Evans, R.M., and Jaenisch, R., Normal development and growth of mice carrying a targeted disruption of the α1 retinoic acid receptor gene., *Proc. Natl. Acad. Sci. U.S.A.*, 90, 1590, 1993.

65. Dickson, C., Smith, R., Brookes, S., and Peters, G., Tumorigenesis by mouse mammary tumor virus: proviral activation of a cellular gene in common integration region int-2, *Cell*, 37, 529, 1984.

66. Moore, R., Casey, G., Brookes, S., Dixon, M., Peters, G., and Dickson, C., Sequence, topography and protein coding potential of mouse int-2: a putative oncogene activated by mouse mammary tumour virus, *EMBO J.*, 5, 919, 1986.

67. Klagsbrun, M., The fibroblast growth factor family: structural and biological properties, *Prog. Growth Factor Res.,* 1, 207, 1990.

68. Dickson, C., et al., Characterization of int-2: a member of the fibroblast growth factor family, *J. Cell Sci. Suppl.*, 13, 87, 1990.

69. Smith, R., Peters, G., and Dickson, C., Multiple RNAs expressed from int-2 gene in mouse embryonal carcinoma cell lines encode a protein with homology to fibroblast growth factors, *EMBO J.*, 7, 1013, 1988.

70. Mansour, S.L. and Martin, G.R., Four classes of mRNA are expressed from the mouse int-2 gene, a member of the FGF gene family, *EMBO J.*, 7, 2035, 1988.

71. Wilkinson, D.G., Bhatt, S., and McMahon, A.P., Expression pattern of the FGF-related proto-oncogene int-2 suggests multiple roles in fetal development, *Development*, 105, 131, 1989.

72. Wilkinson, D.G., Peters, G., Dickson, C., and McMahon, A.P., Expression of the FGF-related proto-oncogene int-2 during gastrulation and neurulation in the mouse, *EMBO J.*, 7, 691, 1988.

73. Represa, J., Leon, Y., Miner, C., and Giraldez, F., The int-2 proto-oncogene is responsible for induction of the inner ear, *Nature*, 353, 561, 1991.

74. Mansour, S.L., Thomas, K.R., and Capecchi, M.R., Disruption of the proto-oncogene int-2 in mouse embryo-derived stem cells: a general strategy for targeting mutations to non-selectable genes, *Nature*, 336, 348, 1988.

75. Goldfarb, M., The fibroblast growth factor family, *Cell Growth Differen.,* 1, 439, 1990.

76. Scriver, C.R., Beaudet, A.L., Sly, W.S., and Valle, D., Eds., *The Metabolic Basis of Inherited Disease*, McGraw-Hill, New York, 1989.

77. Tybulewicz, V.L.J., et al., Animal model of Gaucher's disease from targeted disruption of the mouse glucocerebrosidase gene, *Nature*, 357, 407, 1992.

78. Kataoka, T., Powers, S., McGill, C., Fasano, O., Strathern, J., Broach, J., and Wigler, M., Genetic analysis of yeast RAS1 and RAS2 genes, *Cell*, 37, 437, 1984.

79. Nasmyth, K.A., FAR-reaching discoveries about regulation of START, *Cell*, 63, 1117, 1990.

80. Lundgren, K., Walworth, N., Booher, R., Dembski, M., Kirschner, M., and Beach, D., Mikl and weel cooperate in the inhibitory tyrosine phosphorylation of cdc2, *Cell*, 64, 1111, 1991.

81. Hoffman, F.M., *Drosophila* abl and genetic redundancy in signal transduction, *Trends Genet.,* 7, 351, 1991.

82. Ferguson, E.L. and Horvitz, H.R., The multivulval phenotype of certain *Caenorhabditis elegans* mutants results from defects in two functionally redundant pathways, *Genetics*, 123, 109, 1989.

83. Last, R.L., Bissinger, P.H., Mahoney, D.J., Radwanski, E.R., and Fink, G.R., Tryptophan mutants in *Arabidopsis*: the consequences of duplicated tryptophan synthase beta genes, *Plant-Cell*, 3, 345, 1991.

84. Saga, Y., Yagi, T., Ikawa, Y., Sakakura, T., and Aizawa, S., Mice develop normally without tenascin, *Genes Dev.,* 6, 1821, 1992.

85. Rudnicki, M.A., Braun, T., Hinuma, S., and Jaenisch, R., Inactivation of MyoD in mice leads to up-regulation of the myogenic HLH gene Myf-5 and results in apparently normal muscle development, *Cell*, 71, 383, 1992.

86. Piedrahita, J.A., Zhang, S.H., Hagman, J.R., Oliver, P.M., and Maeda, N., Generation of mice carring a mutant apolipoprotein E gene inactivated by gene targeting in embryonic stem cells, *Proc. Natl. Acad. Sci. U.S.A.*, 89, 4471, 1992.

87. Hülskamp, M. and Tautz, D., Gap genes and gradients — the logic behind the gaps, *BioEssays*, 13, 261, 1991.

88. Chambon, P. and Manuel, M., Personal communications, Strasbourg, France.

Chapter 6

Animal Models of Human Genetic Disease

Jean-Louis Guénet

Unite de Génétique des Mammifères

Institut Pasteur de Paris

Paris, France

Contents

0-8493-8950-X/95/$0.00+$.50

I. Introduction

Since the discovery by Garrod in 1902 that alkaptonuria was a metabolic disorder with mendelian inheritance, many pathological conditions in humans have been recognized as being more or less the direct consequence of a defective genetic constitution, and it is foreseeable that, as a consequence of the introduction of molecular techniques in human genetics, this trend will accelerate in the forthcoming years. Parallel to this development in the knowledge of human molecular pathology, animals modeling human pathology have also been identified or created. Such animal models can help in the understanding of the pathogenesis of many diseases and will be of paramount importance to test both the efficiency and the absence of harmful side effects of the therapies that will be designed in the years to come to compensate or substitute the defective gene function in humans.

The purpose of this review is to describe how animal models of human genetic diseases are discovered or produced and what their advantages and limitations are. Rather than being exhaustive and up-to-date, which would be extremely difficult if not completely unrealistic given the rapid development of the subject, instead we have tried to be didactic, selecting a few pertinent examples for each class of model.

II. The mouse is a species of choice

Even if, as we shall mention later, it has a few disadvantages, the mouse is by far the mammal of choice for the discovery or production of a particular model. In addition to its small size and legendary prolificacy, the mouse is the species whose genetics is best known. Most of the strains that are used for research result from the systematic mating of closely related progenitors, in general brothers and sisters, and even if such a mating system does not in itself increase the mutation frequency, which in humans is in the range of 10^{-5} to 10^{-6} per locus per gamete, it makes the discovery of recessive mutation more frequent. (Systematically mating brothers to their sisters confers to each new recessive mutation occurring in a progenitor roughly a one out of four chance to be detected during the following 20 generations of inbreeding. Thus, even if the mutation rate is relatively low in the mouse, the great number of genes able to mutate toward a recessive allele and the very large number of pedigrees that are carefully observed worldwide make the discovery of new mutations much less exceptional than in any other species). If we consider that every year mice are bred by millions worldwide, then it is no surprise to learn that hundreds of spontaneous mutations of all kinds have been discovered and accumulated over the last 50 years. The Jackson Laboratory (Bar Harbor, Maine, U.S.A.) is the world's largest repository, with over a hundred mutant strains maintained either in a "breathing" status or as frozen embryos and distributed worldwide.

In addition, we must note that the mouse has also been extensively used by geneticists for the evaluation of the hazards associated with the use of radiation and chemical mutagens of all kinds. This has also contributed to the enrichment of the collection of mutants, and nowadays, some genes possess several mutant alleles with various deleterious consequences.

The mouse is also at its advantage because it shares with humans the privilege of having the most advanced genetic map of all species. That is in part because the mouse is one of the rare species where breeding viable and fertile offspring from interspecific matings is possible, and geneticists can, thus, set up crosses where polymorphisms of all kinds segregate extensively.[1,2,3] The mouse genetic map represents a total haploid length of about 1600 centiMorgans, with around 1800 gene loci mapped on it, and spans over 19 autosomes, an X chromosome, and a Y chromosome. In humans, the density of markers is around the same order of magnitude, and about 800 of the genes mapped can be considered as homologous between humans-mice. This homology has been established based on different criteria: the genes considered to be 'homologous' code for similar products, they have the same function, or they have been assigned to a particular locus using the same molecular probe and, thus, are encoded by similar DNA sequences. When two or more genes of that kind are found to be linked on the same chromosome, a conserved segment results. To date, approximately 900 cM of the mouse map are made of such segments, where both synteny and gene order have been conserved since the divergence of lineages leading to either mouse or humans. Such maps[4,5,6] are invaluable tools when human geneticists wish to list candidate models for a particular human disease and vice versa. The availability of such a map also makes it possible, in many instances, to identify the mice with a defective genotype well before the onset of a particular abnormal feature, using genetic markers closely linked to the mutated alleles and setting up particular crosses.

Finally, another advantage of the mouse for the production of models is associated with the recent isolation of the so-called ES cells. These are cell lines derived from the inner cell mass of blastocysts, which can be cultured *in vitro* for several generations in an undifferentiated status and retain their capacity to colonize the germ-cell lineage of chimeras.[7,8] These cells, as we shall see later, can be manipulated *in vitro*, like ordinary somatic cell lines, and can be used for the addition of exogenous DNA sequences to the mouse genome (transgenesis) or for the selective replacement of a particular sequence by another (gene targeting).

III. Models resulting from spontaneous or induced point mutations

Over 100 mutations producing obesity with or without diabetes, hair loss or change in structure, dwarfism, skeletal defects, eye defects, inner ear defects, anemia, and metabolic, neuromuscular, or immunological disorders with more or less severe phenotypes, etc. have been reported. These have been listed by Green.[31] Regular updating of this list appears in specialized reviews like *Mouse Genome* (Oxford University Press) or *Mammalian Genome* (Springer Verlag).

Many of these mutations have proved interesting for the understanding of the developmental processes operating in mammals, and a few of them (about 50) have been classified as models of human diseases. When the similarities with the human disease extend to the molecular level, the model is called homologous. When, on the contrary, the similarities are limited to a few pathological or biochemical features, the model is called analogous.

Several reviews about the mouse models of human single gene disorders have already been published in recent years.[9,10,11] Here we will report only three examples of the use of such models for the understanding of the homologous human diseases and their potential value for the development of therapies.

Our first example refers to one of the most popular mouse muscular mutants: the X-linked muscular dystrophy *mdx*. This mutation arose spontaneously in the C57BL/10 inbred strain and was accidentally discovered several years ago because hemizygous *mdx*/Y males and homozygous *mdx/mdx* females have elevated plasma levels of muscle creatine kinase and pyruvate kinase.[12] In these animals, no clinical symptoms appear until about 12 months of age, when the affected animals develop mild incoordination. When histological sections are made from homozygous *mdx/mdx* muscles, one can observe dystrophic changes beginning at three weeks, with repeated alternate episodes of muscle fiber necrosis and regeneration. At the molecular level, the *mdx* locus has been recognized as homologous with the Duchenne/Becker (DMD/BMD) muscular dystrophy locus in humans, because mutant mice, like their human counterparts, do not synthesize the protein dystrophin.

In spite of a rather different phenotype, this mutation has proven of great value for the understanding of the human homologous disease, clearly demonstrating that the lack of dystrophin is not, in itself, the primary cause of the human defect, given that affected mice exhibit very little muscular wasting. It has also proven an excellent experimental tool to test the efficiency of adenovirus-mediated gene transfer in the muscle cells of affected mice[13] and to evaluate the fate of grafted normal muscular cells into *mdx/mdx* muscles.

Another example of the use and value of homologous mouse models concerns the hyperphenylalaninaemic or phenylketonuric mouse mutants. Employing the same diagnostic test used in man, the Guthrie test, Bode and his colleagues[14] treated the offspring of male mice with the powerful chemical mutagen Ethyl-Nitroso-Urea (E.N.U) and found the first model of phenylketonuria (P.K.U). Hyperphenylalaninaemia-1, (*hph1*), the mutant which they reported, proved to be deficient in a cofactor of the enzyme phenylalanine hydroxylase, rather than in the enzyme itself, but using a similar protocol, other mutations with the same P.K.U. syndrome were induced,[15] representing other models of the human disease. However, what proved interesting in this latter case was that, by using such a clever experimental protocol, the authors were able to make the complete inventory of all the metabolic defects susceptible to produce phenylketonuria in man. They also found different deleterious alleles at the Phenylalanine hydroxylase (*Pah*) locus resulting in syndromes with various degree of severity, providing, for the first time, a molecular explanation for the phenomenon of expressivity. Here again, all these mutations resulting in phenylketonuria in mice are very useful tools for gene therapy by cell or gene transfer in humans.

Another example of the value of a homologous mouse mutation as a model for human pathology is offered by a defect in the gene encoding homogentisic acid oxidase and producing alkaptonuria both in humans and mice. As we mentioned earlier in this review, alkaptonuria was the first example of a metabolic disease being transmitted as a recessive mendelian character in humans.[16] It is very uncommon except in the Slovak and Dominican Republics, where its prevalence is rather high. It is a painful disease because crystals of homogentisic acid accumulate in the joints. It is very easy to recognize because the urine of the affected patients turns black after a while when in contact with air or with the underwear. Given its scarcity, the gene coding for this metabolic disorder is very difficult to localize on the human genetic map because it takes time to accumulate large-sized informative families. One of my colleagues has recently reported[17] a mouse homologue for this condition, which we easily mapped to mouse chromosome 16 close to the gene coding for somatostatin Smst. We passed this precious information to colleagues in Germany and in the Slovak Republic, who found evidence of linkage in the syntenic region of human chromosome 3.[18] Thanks to this localization, it is now possible to perform a very reliable genetic counseling in humans, while the mouse model may be used to develop therapies of all kinds.

These three examples of models resulting from spontaneous or induced point mutations exemplify how helpful the homologous mouse models of human disorders can be even if, as we shall see later, the pathology often differs in severity in each species. However, these models have a major drawback: being the result of a mutation which is a random and rare event, their discovery depends on chance. In addition, the recognition of their homology with a human disorder sometimes requires a long time. This is not the case with the other two type of models, which we will now discuss: the models resulting from transgenesis and from gene targeting.

IV. Models resulting from transgenesis

Transgenic animals result from the addition of exogenous DNA sequences into the genome. This can be achieved either by mechanical injection of a cloned gene, eventually genetically engineered, into one of the two pronuclei of the fertilized egg or by transfection of embryonic stem cells (ES cells) grown *in vitro* and then reinjected into the blastocel of a developing embryo. Over the last ten years several transgenic strains have been produced, primarily for the study of the regulation of gene transcription, sometimes as a way of modeling human pathologic conditions. It is predictable that many other strains will be produced in the future, not only because new genes will be cloned, but also, perhaps chiefly, because transgenesis is unique in the sense that it allows the addition of genetic traits that do not normally belong to the genome of the species.

Roughly, one can consider that animal models resulting from transgenesis are of three kinds: (1) the first kind results in pathological conditions because a gene of alien origin is translated into a highly cytotoxic product when expressed in the transgenic cells; (2) the second kind of transgenic model results in pathological conditions because a particular gene, the expression of which is precisely regulated in normal individuals, is either overexpressed or completely and permanently switched off or because it is expressed in tissues where it should not or should no longer be; (3) the third kind is the consequence of the addition of normal or defective genetic information of human origin to the genome of the transgenic animal.

To be complete, we should add a fourth class, which is the consequence of the random intercalation of the microinjected DNA sequences into the genome of the host and which results in frameshift mutations. Although this fourth class could be exemplified by a few interesting models, such as some alleles at the limb deformity locus (ld^{Hd}) or at the hot foot locus (ho^{Ph}), we will not consider it further because mutations of that kind, like spontaneous mutations, occur at random.

A. Models resulting from engineered cellular deficits

The genes that are expressed in a tissue specific manner during embryonic development or in the adult are controlled by regulatory sequences which frequently, although not always, are located upstream of the coding regions.

When these cis-acting regulatory sequences are identified, they can be separated from the coding regions they are normally associated to and used in transgenic mice to direct expression of a protein that is not normally encoded in its genome. Such transgenic animals have been designed using tissue-specific regulatory sequences associated to sequences encoding cytotoxic or potentially cytotoxic proteins to program the ablation of specific cell types. Such selective cell killing provides a relatively straightforward method of generating mutant animals that lack specific cell types or even an entire lineage. Several such programed genetic ablations have been made to address specific questions about the origins or the functions of specific cell types either in the developing embryo or in the adult. A few have been used to create mouse models of particular human diseases.[19-20]

The most common strategy makes use of sequences encoding toxic proteins such as the A chains of the diphtheria toxin (DT-A) or of ricin (R-A), which both block protein synthesis. In this case, the cell-killing effect of the toxic protein takes place as soon as the transgene is switched on, and it is absolute.

Another strategy that has been developed is the consequence of the induced intracellular expression of the enzyme thymidine kinase of the herpes simplex virus.[21] This enzyme is not directly toxic to animal cells, but unlike the mammalian thymidine kinase, it can phosphorylate certain nucleoside analogues such as acyclovir or gancyclovir, converting them into toxic drugs for the dividing cells. In this particular case, the cell-killing effect becomes conditional since it depends upon both the expression of the gene coding for thymidine kinase and the administration of nucleoside analogues.

Transgenic mice with various congenital aplasias have been generated over the last few years. Heyman and colleagues,[22] driving the expression of the tk gene of the herpes virus by a hybrid regulatory element consisting of the mouse μ heavy chain enhancer and the κ light-chain promoter, have produced a B-cell deficit of the immune system, while Breitman and Bernstein[20] have generated myelin-deficient transgenic mice, with a phenotype very similar to the shiverer (*shi*) mutation, by selectively killing the oligodendrial cell population.

Transgenic animals resulting from genetic ablation have also been designed for investigating specific patterns of development, for example, of the lens, of the lymphoid tissue with its many cell types, and of the nervous system; and it is possible, at least in theory, to design any type of mouse model with a missing tissue or cell type(s), provided that the necessary cis-acting regulatory sequence is available. In practice, these methods of genetic ablation, particularly the one using the highly toxic DT-A or R-A toxins as cell-killing agents, have a major drawback, which is the consequence of the extreme sensitivity of eukaryotic cells to these toxins. For such genetic ablations to be useful, it is an absolute prerequisite that the regulatory elements be extremely cell specific and not leaky. If this is not the case (if, for example, there is a very small background expression of the transgene in cells which are not targeted), misleading pathological conditions may result. For this reason, proteins exhibiting lower cytotoxic effects are being investigated.

B. Transgenic mice resulting from the abnormal regulation of a particular gene

One of the very first cases of transgenesis reported, the one resulting in a mouse overproducing human growth hormone,[23] is still one of the best examples for this type of transgenic animal. In this case, the expression of the gene coding for the human growth hormone (hGH) was driven by the ubiquitous promoter of the mouse metallothionein gene, and the constitutive and ectopic expression of the hGH encoding gene in the transgenic mice resulted in a dramatic increase in size associated with several other minor pathological effects.

Several similar models have been generated by disrupting the normal control of gene expression. That is the case for the strains where the transgene is constructed with the coding sequence of a cellular oncogene coupled to a ubiquitous promoter. Mice of that kind develop neoplasias with a very high frequency. When, on the contrary, the promoter is tissue specific, then cancers occur in specific tissue.[24,25]

Producing transgenic mice with abnormally regulated cellular oncogenes is one of the best ways to analyze the mechanisms of oncogenesis because it does not require that a reciprocal translocation occur to activate the oncogene in question. For example, making transgenic mice with a chimerical DNA stretch resulting from the *in vitro* adjunction of the first 5′ exon of the *bcr* gene to the 3′ exons of the c-Abelson oncogene, Heisterkamp et al.[26] have produced a model of the acute human leukemia, which results from the 9q34-22q11 reciprocal translocation producing the so-called Philadelphia chromosome. In so doing they have unravelled the causal relationship between the presence of the Philadelphia chromosome and the development of the acute leukemia. The model they have produced, unfortunately, is not very useful for the study of the evolution of human leukemia because the mice die at a very early age.

C. Transgenic models resulting from the addition of a human gene

1. The case for poliovirus-sensitive transgenic mice

The poliovirus, the causative agent of poliomyelitis, normally infects only in primates, but type 2 virulent strains can infect mice. Transgenic animals susceptible to all three poliovirus serotypes have been produced by injecting *in ovo* the human gene encoding the cellular receptors for the virus.[27] When inoculated, these transgenic mice mimic clinical symptoms similar to those observed in humans and monkeys and, thus, represent excellent models to study the molecular mechanisms of pathogenesis of the poliovirus and to test vaccines against poliovirus infections.

Attempts have also been made to produce similar transgenic mice sensitive to infections with the different types of HIV, the retrovirus responsible for

AIDS, or to HBV, the virus responsible for B-type hepatitis in human beings. So far, they remain unsuccessful. In the case of HBV, however, transgenic mice resulting from the integration of a complete or nearly complete viral genome have permitted the analysis and study of the liver-specific activity of the viral genomes, and they provide a model to analyze some aspects of the pathology linked to HBV chronic carriage.

2. A mouse model for osteogenesis imperfecta type II

A mouse model of the human disease *osteogenesis imperfecta* type II has been produced by engineering, *in vitro*, an abnormal mouse pro-α1 (I) collagen gene homologous to the abnormal human gene, then putting it back into a transgenic mouse.[28] The animals carrying such a transgene appeared very sick soon after birth because of the modification of the extracellular matrix by the abnormal collagen fibers. However this model, having a dominant effect, was difficult to propagate.

3. A mouse model for sickle cell anemia

A transgenic mouse line has been successfully produced by Ryan et al.[29] by insertion of the normal human α globin gene jointly with the abnormal β^s globin gene immediately downstream of the erythroid-specific DNase I super hypersensitive sites normally located 50 kb upstream of the human β-globin gene. The β^s gene is characteristic of sickle cell anemia in humans. When erythrocytes from these transgenic mice were deoxygenated, greater than 90% of the cells displayed the same characteristic sickled shapes as erythrocytes from humans with sickle cell disease. Mice also had decreased hematocrits, lower hemoglobin concentrations, splenomegaly, and anemia symptoms that are associated with human sickle cell disease. Models like this may result in the development of new drugs and therapies for treating such a debilitating disease.

The possibility of introducing native or genetically engineered exogenous DNA sequences into the mouse genome represents an important breakthrough in mammalian genetics. It has permitted, for example, the identification of the sequences involved in gene regulation and, as we mentioned, it has also permitted the production of a new kind of animal model. However, in spite of considerable advantages, this technique has some drawbacks. One of the most significant is that the genetic information encoded by the exogenous DNA sequence is added to (rather than substituted for) the background information contained in the genome of the species. This means that, if we exclude the exceptional cases where the transgene accidentally disrupts a coding sequence, it is not possible to produce recessive alteration by transgenesis. Another drawback is that the injected DNA sequence, as a rule, inserts randomly in the genome of the host and frequently in the form of several copies associated in tandem. Such drawbacks do not apply to the technology that makes use of embryo-derived stem cells (ES cells) for germ line manipulation, which appears to be extremely promising for the production of gene-targeted mutations. This will be the subject of the next section.

D. Models resulting from *in vitro* modifications in ES cells

1. Establishing ES cell lines

Embryonic stem cells (abbreviated ES cells hereafter) are derived from the inner cell mass of mouse blastocysts.[7] To prevent them from differentiating, they are cultured *in vitro* on a feeder layer of fibroblasts in the presence of low concentrations of Leukemia Inhibitory Factor or LIF, and they are transplanted at a relatively rapid pace. These embryonic cells represent the material of choice for the geneticist because they can be manipulated just like ordinary somatic cells as long as they are *in vitro,* but retain their full developmental potential when injected into a normal blastocyst and the blastocyst is placed into the genital tract of a pseudopregnant mouse. Provided that they stay karyotypically normal and that they have not been maintained *in vitro* for too long, ES cells are frequently capable of the germ-cell lineage of mouse chimeras.[8] It is then possible to use the techniques of somatic cell genetics to isolate ES cells carrying a particular type of mutation and introduce these mutations into the germ line.

2. Mouse models resulting from mutant ES cells selected in vitro

A mouse model of the Lesch-Nyhan syndrome, a severe X-linked metabolic disease in humans resulting from the absence of the enzyme hypoxanthine phosphoribosyl transferase or HPRT, has been produced by recycling *in vitro*-selected ES cells in the germ line of mouse chimera HPRT[30]. The strategy that has been used for the production of such a mutation might be applied, at least in theory, to other genes as well. This would require, however, that the product of the targeted gene be expressed *in vitro* and that an efficient protocol exist to select the mutant cells. Given the extremely low frequency of mutation occurring in mammalian cells grown *in vitro*, the technique is of limited value for the genes that are carried by the autosomes because it would also require that the two copies of the gene be mutated simultaneously for the selection to be efficient.

3. Mouse models resulting from in vitro *transgenesis*

Transgenic mice are relatively easy to produce, and several commercial firms are now able to make such animals virtually upon request as soon as the DNA molecule is available. However, as we already mentioned, there is no way to control the site of integration or the number of copies integrated into the genome when transgenic animals are made by *in ovo* microinjection. Using ES cells is advantageous from this point of view because it is possible to use either a variety of retroviral vectors or *in vitro* transfection of DNA molecules followed by selection to obtain transgenics of a particular type. This technique also has the tremendous advantage of reducing the number of mice to be bred when a particular model is searched.

Another mouse model of the Lesch-Nyhan syndrome produced by infecting ES cells *in vitro* with defective retroviruses followed by selection for an HPRT$^-$ phenotype[31] exemplifies this strategy, and many other transgenic animals of that kind have been produced in that manner. Few of them, however, have resulted in a model for a particular human genetic disorder.

Finally, even if it is not the aim of this review to cover in detail the different strategies used for producing modifications of the mammalian genome, it is worth mentioning here that the *in vitro* transfection of ES cells by genetically engineered DNA sequences, such as, for example, the gene of *E. coli* coding for β-galactosidase, is one of the basic approaches used for the identification of new, developmentally significant genes in mammals (gene trapping). By allowing the production of chimeras with a normal counterpart, *in vitro* transgenesis is also the only way to investigate the developmental effects of a transgene when the latter has a dominant lethal effect.

4. Mouse models resulting from in vitro *homologous recombination*

By far the most interesting and also the most promising application of ES cells technology for the production of mouse mutations is the one making use of *in vitro* homologous recombination. As a rule it consists of four steps: (1) a recombinant DNA molecule is engineered with the sequence of the targeted gene as a basic material; (2) the engineered DNA molecules are transfected into the ES cells and only those cells that have been successfully transfected are selected and retained for the next step; (3) only those cells that have homologously exchanged their native copy of the targeted gene for the mutated one are selected; and (4) a few mutated cells are placed into a host blastocyst and the latter is reimplanted into a pseudopregnant mother.

The first endogenous gene to be modified by homologous recombination with a targeting vector was the human β-globin gene,[32] and since this first experiment, several other homologous recombinations have been achieved, some of them leading to the production of invaluable animal models. To exemplify the strategy used in these experiments, we will report with some detail the production of a mouse model for the human cystic fibrosis.

Cystic fibrosis (CF) is a painful disease, which is reputedly the most common lethal hereditary disease transmitted as a mendelian recessive trait in humans. Most symptoms of the disease arise because of lost Cl^- conductance across the wet epithelia that line the intestine, lungs, and exocrine ducts. The resulting alteration in salt and water transport leads to blocked ducts and a host of secondary complications. Thanks to the recent developments of efficient adenoviral vector- and liposome-mediated gene transfer, CF now appears a promising candidate for somatic gene therapy, and it is of cardinal importance to possess an animal model allowing the evaluation of the therapeutic effectiveness.

Such a mouse mutation was achieved almost simultaneously in two laboratories[33,34] by "knocking out", in ES cells, the activity of the mouse homologue of the human gene responsible for CF. The gene encoding for the cystic fibrosis membrane transregulator is localized on mouse chromosome 6 and symbolized *Cftr* in this species. In both cases, the experiment consisted of the *in vitro* alteration of the mouse *Cftr* gene by inserting into one exon of its coding sequence the gene for neomycin phosphotransferase (symbolized *neo*) which, when expressed in mammalian cells, confers the resistance to the antibiotic geneticin or G418. The ES cells successfully transfected with the modified copy of the *Cftr* gene were selected for G418 resistance; then they were cloned and analyzed, first by pools and then individually, in order to identify those that had undergone homologous recombination between the normal endogenous *Cftr* copy and the introduced mutant one. Such a homologous gene substitution occurs at very low frequency, but it is possible to sort out the cells where these rare events occurred using appropriate oligonucleotides as primers for PCR amplification of genomic DNA. If one primer anneals to the inserted exogenous DNA (the *neo* cassette) and the other to the flanking upstream region, then a fragment of predictable size is amplified by PCR. In all other circumstances, there is no product.

The two mouse models of CF (symbolized $Cftr^{-}/Cftr^{-}$) exhibit symptoms that are roughly similar to the symptoms observed in affected children, and in most cases, they die shortly after birth.

The strategy used for the production of mouse models of CF is applicable to all cases where a genetic disease is identified at the molecular level and the mouse homologous gene is cloned. Mice with sickle cell anemia or various thalassemias have been produced, and it is likely that several other similar models will be produced in the years to come for the most prevalent human disorders (Tay-Sachs, Gaucher, etc.).

Aside from genes being "knocked out" by homologous recombination, gene targeting strategies are also actively designed to produce *in vitro* missense and nonsense mutations or, in a more general way, to produce a number of different mutations in a single gene,[35] resulting in a reduced but not null gene activity,[36,37] In these latter cases, the strategy consists of two steps. Insertion vectors are constructed that contain a homology region carrying the desired mutation(s) as well as positively and negatively selectable markers. The vector is first introduced into the target gene through the homologous region it contains. As a result, the homologous region becomes duplicated: one copy provided by the vector and the other by the target gene. Transfected clones are positively selected for through the activity of the positive selectable marker contained in the vector. Subsequently, in some of these clones, a second homologous recombination event eventually takes place between both copies of the homologous region. Such a recombination results in the loss of the entire vector (one copy of the homologous region included). If the copy that is left comes from the original vector (that is, bears the mutation to be introduced) then the resulting clone will have been mutated at the target locus. Clones that have undergone the second recombination event can also be positively selected because they have lost the negative selectable gene carried by the vector.

Finally, strategies are also developed with the aim of producing mutations of both alleles of an autosomal gene. However, this may not be a considerable advantage if the mutated allele has deleterious effects.

V. The value and limitations of mouse mutants as models for human diseases

As more homologous models are identified, it becomes apparent that there are frequently important differences in severity between diseases in humans and in mice. The severity of the human dystrophic syndrome, for example, with its ruthless consequences for the affected boys, contrasts with the relative weakness of the mouse syndrome. The complete absence of ochronosis in alkaptonuric mice was an unexpected discovery given that it is one of the major pathologic features for the homologous syndrome in humans.

To explain these discrepancies, it must be mentioned first that observing differences between mice and humans in terms of severity for a particular genetic disorder can in no way be considered a surprise, given that intraspecific variations are already fairly common. A few years ago, checking the offspring of mutagenized mice for the possibility of induced mutations in the genes coding for the immunoglobulin light and heavy chains, one of my co-workers found an almost perfect model of the human erythropoietic protoporphyria. After some investigation, this mutation was recognized as the consequence of a point mutation in the gene coding for the enzyme ferrochelatase: the enzyme that is in charge of putting an iron molecule inside of the heme molecule. This mutation was symbolized $Fech^{m1Pas}$ and localized on the telomeric part of mouse chromosome 18 in a region that is homologous with human chromosome 18. On the BALB/c inbred background (the background where the mutation was first discovered), $Fech^{m1Pas}/Fech^{m1Pas}$ homozygous mice are icteric, are extremely sensitive to light, and develop a severe, sometimes fatal, cirrhosis. However, backcrossing the same mutation Fechm1Pas onto a C57BL/6 inbred background, my co-worker came to a point where he could hardly recognize the affected animals due to a much weaker syndrome. Examples of that kind are extremely common in the mouse, where it has also been reported that the same mutant allele could be dominant in a particular strain and recessive in another.

Aside from the role of these non-epistatic interactions in the expressivity of a given phenotype, it is likely that the nature of the mutation itself, considered in molecular terms, is also important. In humans, for example, at least 230 different point mutations have been identified at the CFTR locus. Some of them produce the classical syndrome of cystic fibrosis, with its lethal consequences at the level of the respiratory and intestinal tracts; others, on the contrary, are detected only when human males affected by complete azoospermia are checked at this locus. It makes sense that when a mutation occurs in a particular gene, the severity of the phenotype depends upon what remains in terms of enzymatic activity. A null allele may generate a very severe syndrome, leading to death at an early stage of development, while as low as a few percent activity may be sufficient for the individual to survive and exhibit a more or less severe pathological syndrome. In a similar way, a missense mutation resulting in an abnormal product may have a dominant effect because the abnormal product in question interacts with the product(s) of the normal gene copy(ies) to produce an abnormal protein. With the increasing number of genes being "knocked out", which means changed for a null allele, we should learn much more about such subtle gene interactions.

In addition to the points we just mentioned, Erickson[38] suggested three other possible mechanisms to account for the discrepancies between human and mouse syndromes for homologous genetic disorders. These are (1) variations in biochemical pathways between mouse and humans; (2) variations in developmental pathways; and (3) absolute time vs. physiological time for the rate of development of a pathological process.

To exemplify his first point, Erickson suggests that the Lesch-Nyhan syndrome, which is a very severe disease in humans, has no equivalent in *Hprt⁻/Hprt⁻* mice because mice, but not humans, can convert uric acid to allantoin by oxidation with urate oxidase and are, thus, protected from the potentially toxic metabolites of uric acid. Similarly, it is possible that the alkaptonuric mouse model does not develop ochronosis simply because mice cure this pathological trait by overproducing Vitamin C in their caecum.[17]

To illustrate his second point, Erickson notes that the relative difference in size between mice and humans may influence some developmental processes. Humans deficient in carbonic anhydrase (CAII) suffer from osteopetrosis, renal tubular acidosis, intracranial calcifications, and mental retardation, while mice affected at the homologous locus never develop osteopetrosis. According to Erickson, this may be the consequence of the fact that rodents remodel their smaller bones only at their surfaces and not internally, as larger mammals do. Likewise, it is also accepted now that some of the human thalassemias will never be modeled by a simple mouse mutation because mice, unlike humans, have no fetal hemoglobin.

Erickson finally considers that there is a more or less time-dependent rate for the onset of symptoms encoded for by homologous defective genes. This may be true, for example, when a toxic metabolite is produced at constant rate and accumulates progressively in the organism, producing a pathological syndrome only in the old subjects.

It is likely that the reasons mouse mutants differ from their human counterparts are both numerous and strongly interacting during the development of the mutant organism, and we probably still have a lot to learn in the matter. There is no doubt, however, that the research will yield beneficial knowledge, because what is important, after all, is not to note that a mouse can behave normally with a complete lack of dystrophin, but to understand why and how.

VI. Conclusion

The collection of mouse mutations which have been accumulated over the last six decades by the former geneticists represent an invaluable source of animal models for the study of homologous human genetic disorders. Some of them are now known in molecular terms and employed to design and test therapies of potential use for humans. It is likely that in the forthcoming years, these trends will accelerate because, whereas until recently the discovery of desired mutants was exclusively a matter of chance, techniques have now been developed that make it possible to tailor animals with predetermined alterations in their genome. In this respect, it is not unrealistic to expect that mouse genetics, and with it mammalian genetics, will undergo a revolution.

References

1. Guenet, J.-L., Simon-Chazottes, D. and Avner, P.R., The use of interspecific mouse crosses for gene localization: present status and future perspectives, *Curr. Top. Microbiol. Immunol.,* 137, 13-17, 1988.

2. Avner, P.R., Amar, L., Dandolo, L. and Guenet, J.-L., Genetic analysis of the mouse using interspecific crosses, *Trends Genet.,* 4, 18-23, 1988.

3. Copeland, N.G. and Jenkins, N.A., Development and applications of a molecular genetic linkage map of the mouse genome, *Trends Genet.,* 7, 113-118, 1991.

4. Nadeau, J.H., Maps of linkage and synteny homologies between mouse and man, *Trends Genet.*, 5, 82-86, 1989.

5. Lyon, M.F. and Kirby, M.C., Mouse chromosome atlas, *Mouse Genome*, 91(1), 40-80, 1993.

6. O'Brien, S.J., Womack, J.E., Lyons, L.E., Moore, K.J., Jenkins, N.A. and Copeland, N.G., Anchored reference loci for comparative genome mapping in mammals, *Nat. Genet.,* 3, 103-112, 1993.

7. Evans, M.J. and Kaufman, M.H., Establishment in culture of pluripotential cells from mouse embryos, *Nature*, 292, 154-156, 1981.

8. Bradley, A., Evans, M.J., Kaufman, M.H. and Robertson, E., Formation of germline chimeras from embryo-derived teratocarcinoma cell lines, *Nature*, 309, 255-256, 1984.

9. Leiter, E.H., Beamer, W.G., Shultz, L.D., Barker, J.E. and Lane, P.W., Mouse models of genetic diseases, in *Medical and Experimental Mammalian Genetics: A Perspective*, McKusick, V.A., Roderick, T.H., Mori, J. and Paul, N.T., Eds, Alan R. Liss, New York, 1987, 221-257.

10. Winter, R.M., Malformation syndromes: a review of mouse/human homology. *J. Med. Genet.,* 25, 480-487, 1988.

11. Darling, S.M. and Abbott, C.M., Mouse models of human single gene disorders I: non transgenic mice, *BioEssays*, 14, 359-366, 1992.

12. Bulfield, G., Siller, W.G., Wight, P.A.L. and Moore, K.J., X chromosome-linked muscular dystrophy (*mdx*) in the mouse. *Proc. Natl. Acad. Sci. U.S.A .,* 81, 1189-1192, 1984.

13. Ragot, T., Vincent, N., Chafey, P., Vigne, E., Gilgenkrantz, H., Conton, D., Cartaud, J., Briand, P., Kaplan, J.C. and Perricaudet, M., Efficient adenovirus-mediated transfer of human minidystrophin gene to skeletal muscle of *mdx* mice, *Nature*, 361, 647-650, 1993.

14. Bode, V.C., McDonald, J.D., Guenet, J.-L. and Simon, D., *Hph-1*, a mouse mutant with hereditary hyperphenylalaninemia induced by ethyl-nitroso-urea mutagenesis, *Genetics*, 118, 299-305,1988.

15. McDonald, J.D., Bode, V.C., Dove, W.F. and Shedlovsky, A., Pah^{hph-5}: a mouse mutant deficient in phenylalanine hydroxylase, *Proc. Natl. Acad. Sci. U.S.A.,* 87, 1965-1967, 1990.

16. Garrod, A.E., The incidence of alkaptonuria: a study in chemical individuality, *Lancet II*, 2, 1616-1620, 1902.

17. Montagutelli, X., Lallouette, A., Coud,, M., Kamoun, P., Forest, M. and Guenet, J.-L., aku, a mutation of the mouse homologous to human alkaptonuria, maps to chromosome 16, *Genomics,* 19. 9-11, 1994.

18. Janocha, S., Wolz, W., Srsen, S., Srsnova, K., Montagutelli, X., Guenet, J. -L., Grimm, T., Kress, W. and Müller, C.R., The human gene for alkaptonuria (aku) maps to chromosome 3q, *Genomics,* 19, 5-8, 1994.

19. Bernstein, A. and Breitman, M.L., Genetic ablation in transgenic mice, *Mol. Biol. Med.*, 6, 523-530, 1989.

20. Breitman, M.L. and Bernstein, A., Engineering cellular deficits in transgenic mice by genetic ablation, in *Transgenic Animals,* Grosveld, F. and Kollias, G., Eds., Academic Press, London, 1992, chap. 6.

21. Borelli, E.R., Heyman, R., Hsi, M. and Evans, R.M., Targeting of an inducible toxic phenotype in animal cells, *Proc. Natl. Acad. Sci. U.S.A.*, 85, 7572-7576, 1988.

22. Heyman, R.A., Borrelli, E., Lesley, J., Anderson, D., Richman, D.D., Baird, S.M., Hyman, R. and Evans, R.M., Thymidine kinase obliteration: creation of transgenic mice with controlled immune deficiency, *Proc. Natl. Acad. Sci. U.S.A.*, 86, 2698-2702, 1989.

23. Palmiter, R.D., Brinster, R.L., Hammer, R.E., Trumbauer, M.E., Rosenfeld, M.G., Birnberg, N.C. and Evans, R.M., Dramatic growth of mice that develop from eggs microinjected with metallothionein-growth hormone fusion genes, *Nature,* 300, 611-615, 1982.

24. Cory, S. and Adams, J.M., Transgenic mice and oncogenesis, *Annu. Rev. Immunol.*, 6, 25-48, 1988.

25. Jenkins, N.A. and Copeland, N.G., Transgenic mice in cancer research, *Import. Adv. Oncol.,* 61-67, 1989.

26. Heisterkamp, N., Jenster, G., ten Hoeve, J., Zovich, D., Pattengale, P.K. and Groffen, J. Acute leukemia in *bcr/abl* transgenic mice, *Nature*, 344, 251-253, 1990.

27. Koike, S., Taya, C., Kurata, T., Abe, S., Ise, I., Yonekawa, H. and Nomoto, A., Transgenic mice susceptible to poliovirus, *Proc. Natl. Acad. Sci. U.S.A.*, 88, 951-955, 1991.

28. Stacey, A.J., Bateman, T., Choi, T., Mascara, T., Cole, W. and Jaenisch, R., Perinatal lethal *osteogenesis imperfecta* in transgenic mice bearing an engineered mutant pro-α1 (I) collagen gene, *Nature*, 332, 131-136, 1988.

29. Ryan, T.H., Townes, T.M., Reilly, M.P., Asakura, T., Palmiter, R.D., Brinster, R.L. and Behringer, R.R., Human sickle hemoglobin in transgenic mice, *Science*, 247, 566-567, 1990.

30. Hooper, M., Hardy, K., Handyside, A., Hunter, S. and Monk, M., HPRT-deficient (Lesch-Nyhan) mouse embryos derived from germline colonization by cultured cells, *Nature*, 326, 292-295, 1987.

31. Kuehn, M.R., Bradley, A., Robertson, E.J. and Evans, M.J., A potential animal model for Lesch-Nyhan syndrome through introduction of HPRT mutations into mice, *Nature*, 326, 295-298, 1987.

32. Smithies, O., Gregg, R.G., Boggs, S.S., Koralewski, M.A. and Kucherlapati, R.S., Insertion of DNA sequence into the human chromosomal beta-globin locus by homologous recombination, *Nature*, 317, 230-234, 1985.

33. Snouwaert, J.N., Brigman, K.K., Latour, A.N., Malouf, N.N., Boucher, R.C., Smithies, O. and Koller, B.H., An animal model for cystic fibrosis made by gene targeting, *Science*, 257, 1083-1088, 1992.

34. Dorin, J.R., Dickinson, P., Alton, E.W.F.W., Smith, S.N., Geddes, D.M., Stevenson, B.J., Kimber, W.L., Fleming, S., Clarke, A.L., Hooper, M.L., Anderson, L., Beddington, R.S.P. and Porteous, D.J., Cystic fibrosis in the mouse by targeted insertional mutagenesis, *Nature*, 359, 211-215, 1992.

35. Evans, M.J., Potential for genetic manipulation of mammals, *Mol. Biol. Med.,* 6, 557-565, 1989.

36. Hasty, P., Ramirez-Solis, R., Krumlauf, R. and Bradley, A., Introduction of a subtle mutation into the *Hox-2.6* locus in embryonic stem-cells, *Nature*, 350, 243-246, 1991.

37. Valancius, V. and Smithies, O., Testing an "in-out" targeting procedure for making subtle genomic modifications in mouse embryonic stem cells, *Mol. Cell. Biol.,* 11, 1402-1408, 1991.

38. Erickson, R.P., Why isn't a mouse more like a man?, *Trends Genet.,* 5, 1-3, 1989.

39. Green, M.C., *Genetic Variants and Strains of the Laboratory Mouse,* 2nd ed., Lyon, M.F. and Searle, A.G., Oxford University Press, New York, 1989.

Chapter 7

Homologous Recombination Proteins and their Potential Applications in Gene Targeting Technology

Stephen C. Kowalczykowski[1] and David A. Zarling[2]

[1]*Division of Biological Sciences*

Sections of Microbiology and of Molecular and Cellular Biology

University of California, Davis

Davis, CA

[2]*Cell and Molecular Biology Laboratory*

SRI International

Menlo Park, CA

Department of Laboratory Medicine

University of California, San Francisco

San Francisco, CA

0-8493-8950-X/95/$0.00+$.50

Contents

I. Introduction

Homologous recombination is the natural means of targeting a gene for replacement by a homologue. By definition, the recombination process is specific, being limited to homologous target sites that exceed a species-imposed minimum length of DNA sequence similarity. Most naturally occurring recombination events take place at the homologous locus, and *in vivo* it is usually efficient. In organisms such as *Escherichia coli* and *Saccharomyces cerevisiae*, foreign DNA introduced into cells by the artificial means of transformation is also integrated homologously. Thus, though seemingly ideal, the use of homologous recombination to target experimental genes in mammalian organisms has met with only partial success. Attempts to homologously target genes in mammalian cells have revealed that the majority of integration events occurs at non-homologous sites.[1-3] To overcome these limitations, researchers in the field of gene therapy, for instance, have turned to relatively higher efficiency, but non-homologous, integration systems employing retroviral or adeno-associated viral vectors to stably introduce DNA into the genome. However, these systems suffer from their own problems. The foremost is that, since non-homologous retroviral recombination is approximately random, the sites of integration cannot be controlled, possibly resulting in the accidental knockout of important functional cellular genes and unpredictable control of target gene expression. Because adeno-associated virus or retroviral vectors cannot be targeted to the homologous locus, gene therapies based on these viral systems cannot correct dominant mutations, and their integration may result in activation of endogenous genes. Retroviral vectors are also limited to use only in replicating cells and they have a tendency to recombine. Finally, target cell specificity is often low in organs receiving the adeno-associated viruses, and retroviruses are immunogenic. Thus, homologous recombination systems possess a number of conceptual advantages.

Unfortunately, the utility of homologous recombination is limited because of its low efficiency, relative to non-homologous recombination, when mammalian cells are transfected with foreign DNA. The reasons for this limitation is not clear and may result from a combination (1) of the complexity of the recombination reaction, (2) of competing processes that attempt to repair (non-homologously) DNA strand breaks, and (3) of potential constraints imposed by nucleosome structure and/or chromatin condensation. It appears that many of the important variables that affect efficient gene targeting are just being described. Rather than discuss potential barriers to efficient homologous recombination in mammalian cells, this article will focus on the biochemistry of the recombination apparatus. Low efficiency DNA transfer and targeting can result from a rate-limiting impediment in any one of a number of key steps in the recombination process. Through an appreciation of the entire process, rational procedures can be designed to overcome the inhibitory process(es) encountered in mammalian cells. Therefore, this article will provide an overview of the salient features of homologous recombination in *E. coli*, the system which is the most biochemically refined. The nature of the enzymatic activities and the substrates that they act upon will be summarized; in addition, the current and future potential application of these enzymes will be highlighted. We hope that appreciation of the enzymology of this process in a relatively simple organism and *in vitro* will guide the design of experiments in more complex and experimentally less tractable systems. For a comprehensive review of these subjects, readers are referred to References 1 through 17.

II. Biochemical activities of key components in homologous recombination

The complexity of homologous recombination in *E. coli* is revealed in Table 1, where the proteins that play a role in recombination are summarized.[10] Over 25 proteins are needed, and their enzymatic activities include DNA strand exchange, DNA renaturation, DNA helicase, nuclease, ATPase, topoisomerase, and DNA-binding activities. Most of these activities are also commonly encountered in other biological processes, but one activity listed in Table 1 is unique to recombination: the DNA strand exchange activity of the recA protein. This activity was anticipated by nearly all models of recombination, but it was surprising that the ability to homologously pair and exchange DNA sequences would reside in a single, relatively small (M_r 38 kDa) polypeptide.

To put the activities of this diverse set of enzymes into context requires appreciation of one key property of recA protein-promoted DNA strand exchange.[4-17] RecA protein will pair two homologous DNA molecules only if one of them is at least partially single-stranded. Since DNA is normally double-stranded, this requirement imposes a necessary initiation step in any recA protein-promoted process: one of the DNA substrates must be processed to reveal ssDNA. This consideration is reflected in Figure 1, which illustrates the plausible function in the recombination mechanism for each of the proteins listed in Table 1. In nearly all models of homologous recombination, the initial event is the generation of a ssDNA or dsDNA break; in *E. coli,* a dsDNA break is created in all of the processes (conjugation, transduction, and transformation) that result in recombination.[18,19] The first step envisioned in Figure 1 is the processing of one of the linear DNA molecules to create ssDNA. This can occur by a variety of biochemical means, but the primary method operative in wild-type *E. coli* is the combined unwinding and degradation of dsDNA by the recBCD enzyme.[10,19,20] This enzyme, in conjunction with the recombination hotspot χ, acts to create ssDNA that can be used by recA protein for the subsequent homologous pairing phase. An alternative approach for the production of ssDNA uses either a strand-specific dsDNA exonuclease or a DNA helicase, or the action of both.

The recA protein, assisted by the ssDNA-binding (SSB) protein (and, perhaps, facilitated by the recF, recO, and recR proteins), promotes invasion of the homologous target. The recA protein is remarkably catholic about the types

Table 1. *Proteins and DNA sites involved in genetic recombination in Escherichia coli* [10]

Protein	Activity
RecA	DNA strand exchange; DNA renaturation; DNA-dependent ATPase; DNA- and ATP-dependent coprotease
RecBCD (exonuclease V)	DNA helicase; ATP-dependent ssDNA and dsDNA exonuclease; ATP-stimulated endonuclease; χ-hotspot recognition
RecBC	DNA helicase
RecE (*sbcA) (exonuclease VIII)**	dsDNA exonuclease, 5′→3′ specific
RecF	ssDNA and dsDNA binding; ATP binding
RecG	Branch migration of Holliday junctions; DNA helicase
RecJ	ssDNA exonuclease, 5′→3′ specific
RecN	Unknown, ATP binding consensus sequence
RecO	Interaction with recR and (possibly) recF proteins
RecQ	DNA helicase
RecR	Interaction with recO and (possibly) recF proteins
RecT	DNA renaturation
RuvA	Holliday-, cruciform-, and 4-way junction binding; interaction with ruvB protein
RuvB	Branch migration of Holliday junctions; DNA helicase; interaction with ruvA protein
RuvC	Holliday junction cleavage; 4-way junction binding
SbcB (exonuclease I) (*xonA*)	ssDNA exonuclease, 3′→5′ specific; deoxyribophosphodiesterase
SbcC	Unknown, ATP binding consensus sequence
SbcD	Unknown
SSB	ssDNA binding
DNA topoisomerase I (*topA*)	ω protein, type I topoisomerase
DNA gyrase (*gyrA* & *gyrB*)	DNA gyrase, type II topoisomerase
DNA ligase (*lig*)	DNA ligase
DNA polymerase I (*polA*)	DNA polymerase; 5′→3′ exonuclease; 3′→5′ exonuclease
Helicase II (*uvrD, uvrE, recL, mutU*)	DNA helicase
Helicase IV (*helD*)	DNA helicase
Chi (χ)	Recombination hotspot: 5′-GCTGGTGG-3′; regulator of recBCD enzyme nuclease activity

From Kowalczykowski, S.C., Dixon, D.A., Eggleston, A.K., Lauder, S.D., and Rehrauer, W.M., *Microbiol. Rev.,* 58, 3, 1994 (in press). With permission.

* *sbcA* mutations are regulatory mutations that activate *RecE* function.

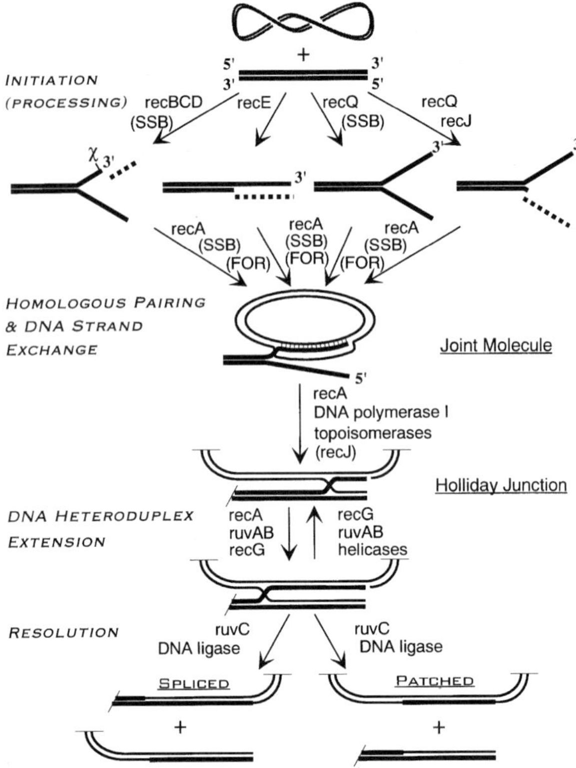

Figure 1. *Biochemical model for homologous recombination in E. coli based on the enzymatic activities of proteins known to be involved based on genetic analysis. (From Kowalczykowski, S.C., Dixon, D.A., Eggleston, A.K., Lauder, S.D., and Rehrauer, W.M.,* Microbiol. Rev., *58, 3, 1994 (in press). With permission.)*

of DNA molecules that it will pair, provided that one of them is at least partially single-stranded. For the reaction illustrated in Figure 1, which is the invasion of a negatively supercoiled DNA recipient by a linear ssDNA fragment, there is a distinct bias for promoting insertion of the 3′-end of the ssDNA.[21] This 3′-end can serve as a primer for any needed repair by replication.

The resulting joint molecule serves as the common intermediate in each of the biochemical pathways. Conversion of this structure to the obligatory Holliday junction, which is common to all mechanisms of reciprocal recombination, could occur by simple pairing of the displaced ssDNA with the other strand of the invading DNA molecule. The Holliday junction is capable of undergoing DNA heteroduplex extension either by a thermally driven branch migration process or by a protein promoted process. Although the thermal process seemingly might suffice for extension through regions of DNA sequence homology, thermally driven extension would terminate at DNA sequence heterologies or by the presence of DNA-binding proteins. Thus, it is not surprising to find a class of proteins that can catalyze this extension process. RecA, ruvAB, and recG proteins all possess this novel activity.[10,12,22]

Finally, resolution of the Holliday junction by symmetric cleavage results in two types of recombinant products: the spliced recombinants, containing the reciprocal halves of the parental molecules, and the patched recombinants, containing nearly intact parental information, but possessing the DNA heteroduplex patch characteristic of a recombination event.[22]

Each of the activities mentioned will be described in more detail. However, only the most important properties of each are described. The reader seeking more in-depth discussion should consult the many fine reviews written on this subject.

A. RecA protein and homologous pairing proteins

The DNA strand exchange activity of *E. coli* recA protein is biochemically unique, yet ubiquitous in nature.[4,5,7,9-11,13-17] All prokaryotes examined possess a gene that bears a high degree of homology to the *E. coli* gene, and this universality appears to extend to eukaryotes as well.[11,15,23,24] RecA protein homologues have been purified from sources as divergent as the closely related organism *Proteus mirabilis*, the gram positive *Bacillus subtilis*, the bacteriophage T4, and the chloroplasts of pea plants.[5,11] Each of these proteins possess the unvarying hallmarks of recA protein-promoted DNA strand exchange: ATP- and ssDNA-dependence. Although ATP-independent homologous pairing proteins have been isolated, these proteins do not pair DNA by the same mechanism employed by recA protein; instead, they require ssDNA in both of the DNA molecules and rely on DNA reannealing as the homology recognition step.[11]

The DNA strand exchange reaction promoted by recA protein is comprised of at least three major phases: presynapsis, synapsis, and DNA heteroduplex extension (Figure 2); each of these major phases can be further dissected into more basic biochemical steps.

The presynaptic step involves the binding and polymerization of recA protein onto the requisite ssDNA. The ssDNA is fully coated by recA protein, requiring about one recA protein monomer for every three to four nucleotides of ssDNA.[25] RecA protein can also bind to dsDNA, but with rather slow kinetics;[26-28] the rate of binding to dsDNA is increased either when ssDNA regions are present[29,30] or when distortions are introduced in the form of either DNA lesions, intercalating dyes, or non-B-DNA character.[31-35] ATP binding, but not hydrolysis, is required for this step,[25,36,37] and the resultant ternary complex of ATP, recA protein, and ssDNA (or dsDNA) forms a distinctive right-handed helical filament.[16] This nucleoprotein filament is about 100 Å in diameter, has a characteristic 95 Å pitch, and extends the DNA by about 50% over the normal dimension of B-form dsDNA. This distinctive ternary complex defines the structure referred to as the presynaptic complex. It is this nucleoprotein filament, rather than a monomer or any other limited aggregate of recA protein, that is the functional entity responsible for the homology search process.

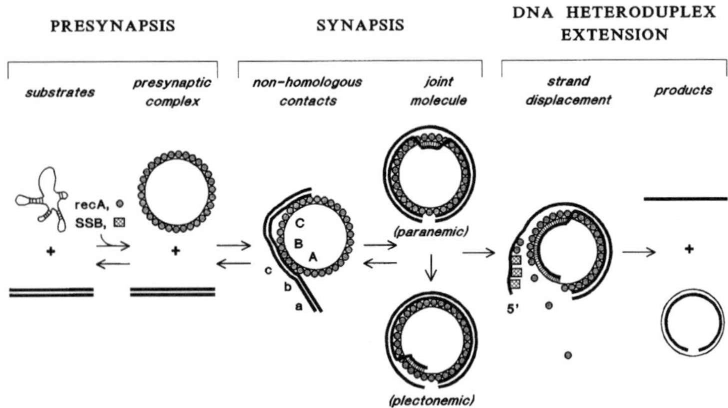

Figure 2. Mechanism of homologous pairing and DNA strand exchange promoted by recA protein. The reaction between circular ssDNA and linear dsDNA is depicted. (From Kowalczykowski, S.C., Annu. Rev. Biophys. Biophys. Chem., *20, 539, 1991. With permission.)*

Once formed, the presynaptic complex is capable of identifying DNA sequence homology in dsDNA. This process is rapid, occurring *in vitro* in just a few minutes; specific, needing less than 50 nucleotide of homology;[38-40] discriminating, rejecting as great as a 200,000-fold excess of non-homologous sequences;[41] and completely independent of ATP hydrolysis.[25,42-47] All of this is made even more remarkable by the fact that the presynaptic filament is 50% more extended than the duplex target homologous sequences.[16] Although many details of the homology search remain unclear, several of the reaction characteristics are known. The search process is first order rather than second order in DNA concentration, implying that the rate-limiting step in the pairing process occurs within non-homologously paired complexes of ssDNA and dsDNA.[48] Homologous recognition does not occur by repeated denaturation of the dsDNA,[49-51] but rather, through transient non-Watson-Crick hydrogen bonding interactions between ssDNA and dsDNA, leading to a 3′-strand-containing structure,[52-54] at least transiently.[44] During the homology search, the dsDNA is topologically unwound, presumably due to distortion required to transiently align the dsDNA with the extended dimensions of the presynaptic filament.[55-59]

Once homologous alignment is achieved, invasion of the dsDNA by the ssDNA and exchange of DNA strands results in formation of a joint molecule or D-loop (displacement loop) structure. Depending on where pairing initiates, two kinds of joint molecules can be detected. If pairing occurs at the ends of a linear molecule, the exchanged DNA strands are free to intertwine, resulting in the formation of a plectonemic joint molecule; if pairing occurs at internal sites where topological constraints limit intertwining, then a paranemic joint molecule forms. Though paranemic joint molecules formed with a single-stranded DNA are unstable in the absence of recA protein, they can convert to plectonemic joint by extending the region of pairing to a free DNA end[43,60] or by the more likely *in vivo* pathway, probably involving a topoisomerase.[56,61,62] As will be discussed in more detail below, protein-free D-loop structures formed at internal dsDNA sites are kinetically unstable in the absence of negative supercoiling due to branch migration; however, the stability of these joint molecules can be increased substantially by hybridization of complementary ssDNA to the displaced strand of the D-loop.[63,64]

The joint molecule is a structure central to the recombination process. The stability of this structure is due to conventional base-pairing. DNA heteroduplex extension, the third phase of this recA protein-promoted process, increases its stability. RecA protein, itself, is capable of promoting DNA heteroduplex extension at a rate of about 10 to 20 bp/sec.[25] This process requires ATP hydrolysis and is unidirectional, occurring $3' \rightarrow 5'$ relative to the ssDNA strand displaced. It can proceed for at least 7 kb and can traverse heterologies in the DNA as large as 1308 nucleotides.[65]

These fundamental properties of DNA strand exchange promoted by recA protein were determined using model DNA substrates (6 to 7 kb) *in vitro* that were totally devoid of protein. However, genomic DNA is clearly coated with polyamines and proteins, among them histones and chromatin-associated proteins in eukaryotic cells and histone-like proteins (e.g., HU) in prokaryotic cells. Not surprisingly, these proteins inhibit the pairing reactions of recA protein.[66-68] The binding of *E. coli* HU protein to dsDNA blocks formation of plectonemic, but not paranemic, joint molecules. Similarly, in a heterologous *in vitro* reaction, eukaryotic chromatin substrates formed by reconstitution of histones with dsDNA were not efficiently utilized by *E. coli* recA protein. Reconstitution at histone/DNA mass ratios of 0.8 and 1.6 suppressed the DNA heteroduplex extension phase, but had no effect on the initial homologous pairing step; however, at a higher mass ratio of 9.0, no pairing was detected. These effects were exacerbated by addition of histone H1 to approximately physiological levels. Thus, the removal of histones and other DNA-binding proteins may represent an important aspect of efficient homologous recombination in the cell. Such clearing may be aided by DNA helicases (see below).

The only other homologous pairing protein described in *E. coli* is the recT protein, a protein encoded by a cryptic lambdoid phage, rac, and, for that reason, bearing functional resemblance to the β-pairing protein from bacteriophage λ.[69] The recT protein binds to ssDNA (but not to dsDNA) and promotes the ATP-independent renaturation of complementary single-stranded DNA.[70] The recT protein can also promote homologous pairing between circular ssDNA and linear dsDNA, provided that the linear dsDNA is resected by an exonuclease to expose homologous ssDNA.[71] In the absence of nuclease activity, no pairing is detected; the recE protein (see below) of *E. coli* can provide this needed nuclease function *in vitro*. This recT protein-promoted pairing reaction also requires no ATP and at least one recT monomer per 13 nucleotides of ssDNA.[71] The pairing reaction likely initiates via reannealing of the homologous ssDNA regions, consistent with the known renaturation activity of recT protein; however, heteroduplex formation is not limited to the region of resected DNA, since ssDNA is displaced from duplex region of the linear dsDNA. This strand displacement or DNA heteroduplex extension is recT protein specific,[71] since a protein proficient in DNA renaturation (histone H1 protein) can promote the initial pairing (reannealing) step, but cannot produce DNA heteroduplex products possessing displaced ssDNA.[71,72] These properties of recT protein resemble those of the ATP-independent class of homologous pairing proteins isolated from many eukaryotic cells,[11] suggesting that ssDNA annealing may be an important pathway for recombination in eukaryotic cells.[3]

B. RecBCD enzyme and DNA helicases

DNA helicases figure prominently in homologous recombination (Table 1). These enzymes can act at two different steps of the recombination process: early in initiation or later in DNA heteroduplex extension.

For the major pathway of recombination in *E. coli*, recBCD enzyme plays a major role as the initiator of DNA strand exchange. The recBCD enzyme is a multi-functional heterotrimeric enzyme encoded by the recB, recC, and recD genes.[10,73,74] It possesses both DNA helicase and nuclease activities and, hence, is unusual among helicases. The nuclease activity is, however, down-regulated by the recombination hotspot χ (see below), resulting in a modified enzyme with enhanced recombination activity that may consist of only the recB and recC subunits, which is primarily a helicase.[75-77]

By means of its helicase activity, the recBCD enzyme is capable of rapidly and processively producing large amounts of ssDNA. Duplex DNA with flush or nearly flush (ssDNA overhangs of less than about 25 nucleotides) dsDNA ends is unwound at about a rate of 1000 bp of dsDNA per second per recBCD enzyme molecule at 37°C.[78,79] Unwinding is highly processive, with an average of about 30 kb unwound before dissociation of the recBCD enzyme.[80,81]

In addition to helicase activity, the recBCD enzyme has a variety of nucleolytic capabilities; these include ATP-dependent dsDNA exonuclease, ATP-dependent ssDNA exonuclease, and ATP-stimulated ssDNA endonuclease activities. Both its helicase and nuclease activities are accommodated by a mechanism for enzyme action that involves degradation of ssDNA produced during DNA unwinding (Figure 3).[80,82-85] Degradation during DNA unwinding is highly asymmetric, with the DNA strand 3′ at the entry site being degraded much more extensively than the strand 5′ at the entry site.[76]

Just as for the recA protein, non-specific DNA-binding proteins impede the action of recBCD enzyme, but in contrast to recA protein, recBCD enzyme can displace these proteins.[86] RecBCD enzyme can displace eukaryotic histones from both reconstituted and native chromatin templates, although these nucleosomal substrates reduce the processivity of unwinding by more than 20-fold and the rate of unwinding by about 3-fold.[86] This *in vitro* finding argues that

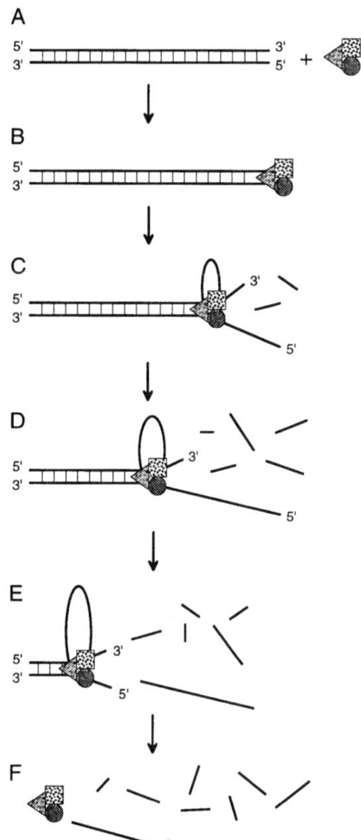

Figure 3. *Model for the mechanism of DNA unwinding and degradation of linear dsDNA by recBCD enzyme. (From Kowalczykowski, S.C.,* Experientia, *50, 204, 1994. With permission.)*

the enzyme should be capable of unwinding prokaryotic chromosomal DNA, which is condensed by histone-like proteins such as HU and IHF to form a "nucleoid" structure in *E. coli* and raises the possibility that specific helicases in eukaryotic cells may serve a similar function.

In addition to recBCD enzyme, there is at least one other DNA helicase that potentially functions at the initiation step of recombination. This is the recQ helicase, and its need in recombination is manifested in cells defective for recBCD function and containing needed suppressor mutations.[87] As for all helicases, hydrolysis of either ATP or dATP is required for dsDNA unwinding by recQ protein.[88] In the absence of SSB protein, the recQ protein is a relatively poor helicase, unwinding only about 143 base pairs and requiring high concentrations of protein. However, SSB protein stimulates its DNA helicase activity, presumably by binding to the unwound strands. Under these conditions, up to 343 base pairs are unwound and unwinding approaches catalytic behavior.[89] Presumably this limited level of DNA unwinding suffices for initiation of DNA strand exchange by recA protein; alternatively, the recQ protein, in concert with a nuclease such as the recJ protein (Figure 1 and see below), can produce longer regions of ssDNA.

In addition to their essential role in initiation of recombination, helicases can also act at the DNA heteroduplex extension step.[12,22] Extension of the region of DNA heteroduplex can stabilize joint molecules by preventing their loss through random branch migration in the reverse direction. Although the recA protein is capable of promoting such a uni-directional reaction, two additional DNA helicases (ruvAB and recG proteins) exist that are both important to recombination and possess the unique attribute of acting on DNA crossover structures (Holliday junctions).[22] The ruvAB protein complex can unwind up to 558 base pairs with an apparent 5′→3′ polarity,[90] but more importantly, it can act on Holliday junctions that form as intermediates in recA protein-promoted DNA strand exchange reactions.[91-95] In contrast to the recA protein-mediated reaction, DNA heteroduplex extension by ruvAB protein is bi-directional[91] and requires as little as one each of the ruvB dimer and ruvA tetramer complex per 500-1200 nucleotides of DNA.[94]

A second DNA helicase, with activity similar to that of ruvAB protein, is the recG protein.[96] The recG protein binds as a monomer to Holliday structures and disrupts the structure in an ATP-dependent manner by virtue of its branch migration activity.[96-98] The recG protein is both more active on a molar basis (ca. 10^3-fold) and faster (ca. 10-fold) than the ruvAB protein in the branch migration reaction.[97] Unlike ruvAB protein, however, the recG protein appears to be specific for crossover structures, since it is unable to unwind conventional helicase substrates.[97]

C. Nucleases

Nucleases also figure prominently in recombination processes. Like DNA helicases, nucleases can function at different steps in the recombination process. In the initiation step, a nuclease acting alone or in concert with other proteins can, in theory, produce ssDNA suitable for use by recA protein. Figure 1 illustrates how two nucleases, the recE protein (exoVIII) and the recJ protein, might serve in that capacity. The other essential purpose for nucleases is resolution of the Holliday junction; this important function is handled by the ruvC protein.

The recE protein is a 96 kDa protein[69] that preferentially degrades the 5′-terminal strand of dsDNA.[99] The protein shows little nuclease activity on ssDNA, but will act on nearly any form of linear dsDNA, except dsDNA containing 5′-ssDNA tails longer than 250 nucleotides.[99] Degradation is highly processive, with about 10,000 to 20,000 nucleotides removed per binding event (Hall and Kolodner, personal communications[152]).

The recJ protein is another recombinationally significant nuclease. In contrast to the recE protein, recJ protein is specific for ssDNA, degrading in the

5′→3′ direction at a maximal rate of about 1000 nucleotides per min per protein molecule.[100] Its precise role in recombination is unknown, but genetic analyses suggest that recJ protein may work in concert with a helicase such as the recBC (i.e., without recD) or recQ helicases,[101] degrading one of the DNA strands produced by DNA unwinding (Figure 1). This degradation may prevent reannealing of the unwound ssDNA strands and has the net effect of producing ssDNA with a 3′-terminal end, which is the preferred substrate for recA protein-promoted invasion of negatively supercoiled DNA.[21]

The last nuclease discussed in this section has the remarkable property of recognizing and symmetrically cleaving Holliday junctions, arguing that this protein catalyzes the last step depicted in Figure 1. This nuclease is the ruvC protein. RuvC is a homodimeric protein composed of subunits with an estimated M_r of 18,747 Daltons.[102] In the absence of Mg^{2+} or ATP, ruvC protein will bind to Holliday junctions, but not cleave the DNA.[103] In the presence of Mg^{2+}, ruvC protein will cleave at the junction, provided that at least six base pairs are fully homologous.[103] Cleavage is symmetric, occurring on the 3′ side of thymine residues, and shows no bias, producing equal amounts of patched and spliced recombinant products.[102,104] This unbiased pattern changes somewhat in the presence of recA protein, with spliced products appearing more frequently than patched products.[103,105] Unexpectedly, the reaction is apparently not catalytic, requiring as much as one ruvC monomer per 80 nucleotides of DNA.[103]

D. SSB protein

DNA strand exchange promoted by recA protein is stimulated by the SSB protein. SSB protein is a representative of a class of ssDNA-binding proteins that has no enzymatic activity, but binds ssDNA cooperatively and non-specifically. SSB protein (M_r18.8 kDa[106]) binds both single-stranded DNA and RNA, resulting in a destabilization (i.e., lowering of the T_m) of duplex nucleic acid structures.[107]

The stimulatory effects of SSB protein are manifested both pre- and post-synaptically (Figure 4). In the presynaptic phase, SSB protein, by virtue of its helix-destabilization properties, removes the structural impediment to complete presynaptic complex formation by removing DNA secondary structure.[108,109] The amount of SSB protein required for optimal presynaptic complex formation is about 1 SSB monomer per about 15 to 20 nucleotides of ssDNA.

Post-synaptically, SSB protein serves two functions. The first is to prevent formation of homologously paired networks of DNA that result from intermolecular re-invasion events; SSB protein can bind to the displaced linear ssDNA and hinder its utilization by recA protein.[110] The second role is more direct, requiring sufficient SSB protein to bind the ssDNA produced by DNA strand exchange; the binding of SSB protein to the displaced ssDNA directly stimulated the observed rate of joint molecule formation, presumably because it limits reversal of DNA strand exchange.[111]

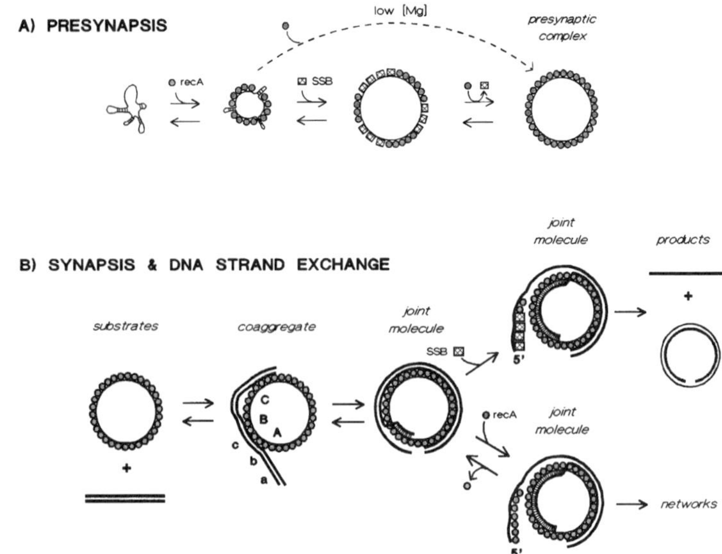

Figure 4. *Role of SSB protein in recA protein-promoted DNA strand exchange: presynaptic (**A**) and synaptic (**B**) functions. (From Lavery, P.E., and Kowalczykowski, S.C.,* J. Biol. Chem., *267, 9315, 1992. With permission.)*

E. Recombination hotspots

The frequency of recombination is not uniform across the genome, but certain regions have increased probabilities of crossing-over.[112] These sites are referred to as recombination hotspots, and *E. coli* possesses a specific hotspot known as χ. Though initially identified in bacteriophage λ, χ sites are believed to be responsible for most, if not all, recombination in *E. coli*. Thus, although initially identified as a specialized site of recombination, it appears as though all recombination in wild-type cells is mediated by χ sites, and therefore, they are general rather than a specialized feature of recombination.

Recombination is 5 to 10 times more frequent near χ; is dependent on *recB* and *recC*; is stimulated primarily to the 5′ side of the χ site; and is detectable as far away as 10 kb.[113-116] The DNA sequence 5′-GCTGGTGG-3′ constitutes the χ site[117]. *In vitro*, recBCD enzyme recognizes χ only when unwinding DNA from the 3′-side of the χ sequence, and recognition of χ produces a ssDNA fragment that terminates near χ; cleavage occurs in the DNA strand containing the χ sequence, 4 to 6 nucleotides to the 3′ side of χ.[118,119]

III. Coordination of biochemical functions *in vitro*

The detailed biochemical characterization of many of the proteins listed in Table 1 has permitted limited reconstitution of the recombination process. In one reaction, the initial steps of homologous pairing depicted in Figure 1 have been reconstituted *in vitro*. In this reaction, homologous pairing is dependent on the concerted action of recA, recBCD, and SSB proteins. In addition, homologous pairing in this reaction is stimulated by the presence of the recombination hotspot χ. In a second reaction, Holliday junctions are formed *in vitro* by recA protein and are specifically resolved by the ruvC protein to produce the expected recombinant progeny.

A. Homologous pairing dependent on recA, recBCD, and SSB proteins and χ

The first *in vitro* experiments were designed to test whether recBCD enzyme could generate ssDNA that was a suitable substrate for recA protein-dependent homologous pairing reactions.[120,121] These reactions used DNA molecules that were not substrates for recA protein; however, one of the participants was a linear DNA molecule that could be unwound by the helicase activity of recBCD enzyme. These recombination reactions showed that the recBCD enzyme could, indeed, initiate recombinational events by producing ssDNA suitable for recA protein-dependent DNA heteroduplex formation,[120-123] demonstrating *in vitro* that recA, recBCD, and SSB proteins were sufficient to reconstitute the first two steps depicted in Figure 1.

Successful reconstitution of the initiation phase of the recombination reaction permitted an evaluation of the effect of the recombination hotspot χ on the joint molecules formed *in vitro*. The nick-initiation model of Smith and colleagues[124] predicted the formation of joint molecules containing ssDNA originating from the donor DNA (i.e., the dsDNA unwound by recBCD enzyme and the source of ssDNA for subsequent strand invasion) and derived from DNA downstream of χ. *In vitro*, the prediction was not only verified, but DNA heteroduplex formation was significantly enhanced when χ-containing DNA was used as the donor DNA.[75] In addition, these reactions revealed an unanticipated feature of χ: the χ recombination hotspot is also a regulatory sequence. It effects a change in the enzymatic behavior of recBCD enzyme, converting this multifunctional enzyme from a combined helicase-nuclease to a modified enzyme possessing essentially only helicase activity.[75,76] Interaction with χ attenuates the nuclease activity of recBCD enzyme, an effect that lasts as long as the enzyme is associated with the DNA molecule that contained χ. Upon dissociation, nuclease function is normally restored and the cycle of attenuation and reactivation can continue indefinitely. These observations provide a simple rationale for the recombination hotspot activity of χ. In the absence of χ, recBCD enzyme degrades and destroys the ssDNA needed for homologous pairing by recA protein. However, upon recognition of χ, this needed ssDNA is spared from subsequent degradation, allowing an increased probability of a productive pairing event.

The precise nature of the molecular event responsible for nuclease attenuation is unclear, but both *in vivo* and *in vitro* experiments argue for loss or modification of the *recD* subunit as the cause.[77,125,126] *In vivo*, both the phenotype and distribution of crosses of certain *recD* mutants imitate the behavior of wild-type, χ-stimulated, *recBCD*-dependent recombination.[125,126] *In vitro*, the recBC enzyme possesses characteristics similar to the recBCD enzyme that has interacted with χ; most notably, the recBC enzyme has essentially no nuclease activity, but possesses significant helicase activity.[77,127] Thus, the recBC enzyme most likely represents the modified enzyme that is constitutively proficient in recombination.[77,125]

The nuclease attenuation model for initiation of genetic recombination illustrated in Figure 5 summarizes the salient features of the *in vitro* data.[75,76] The recombination reactions initiate by the unwinding and degradation of dsDNA by recBCD enzyme. The combined unwinding and nucleolytic action produce ssDNA fragments that are adequate substrates for recA protein-dependent reactions *in vitro,* but do not provide the continuity needed to produce crossover recombinants *in vivo*. RecBCD enzyme degrades the dsDNA asymmetrically, with the DNA strand 3′ at the entry site being nicked more frequently than the 5′ strand. Unwinding and degradation continue up to χ. Interaction with χ causes the recBCD enzyme to pause, securing a high

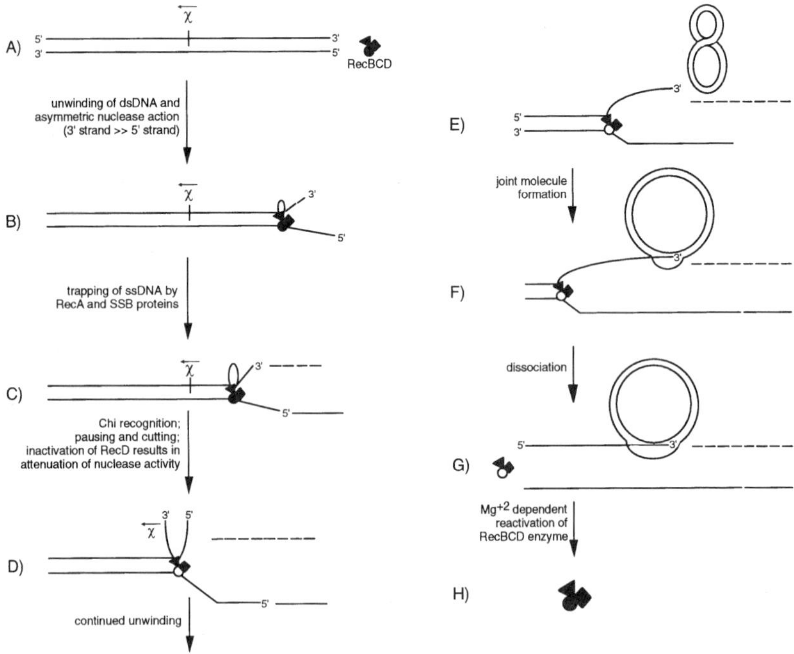

Figure 5. *Nuclease attenuation model for initiation of recombination by recBCD enzyme, χ sites, and both recA and SSB proteins. (From Kowalczykowski, S.C., Dixon, D.A., Eggleston, A.K., Lauder, S.D., and Rehrauer, W.M.,* Microbiol. Rev., *58, 3, 1994 (in press). With permission.)*

probability of an endonucleolytic cleavage in the vicinity of χ. Recognition of χ causes modification or loss of the recD subunit, resulting in at least a 500-fold reduction of the nuclease activity without loss of the helicase activity.[75,76] The net effect is the creation of an enzyme that is primarily a helicase. Without its nuclease activity, recBCD enzyme continues to unwind the dsDNA, producing an intact ssDNA fragment downstream of χ. This ssDNA strand is the substrate for recA protein-promoted DNA strand invasion of the supercoiled DNA recipient (chromosome). The nuclease function is attenuated as long as the χ-containing DNA that produced the attenuation. This attenuation is fully reversible and catalytic because when the modified enzyme dissociates from the DNA, its nuclease activity is reactivated.[75,76]

B. Resolution of Holliday junctions

A second reconstituted *in vitro* reaction reproduces the Holliday junction formation, DNA heteroduplex extension, and resolution steps of the *in vivo* process (Figure 1).[22] This reaction bypasses the initiation step by using processed DNA substrates suitable for the DNA strand exchange activity of recA protein.[128] Using circular dsDNA with a ssDNA gap and homologous linear dsDNA, recA protein can catalyze formation of a Holliday junction, and by virtue of its DNA heteroduplex extension activity, it can promote a protein-dependent migration of the crossover point.[129]

The ruvC protein can resolve the Holliday structure created by recA protein *in vitro* to yield both kinds of recombinant progeny (both sliced and patched molecules).[103,105] On DNA strand exchange intermediates (i.e., Holliday junctions) stripped of recA protein, production of either spliced or patched products is unbiased; however, in the presence of recA protein, production is biased towards formation of splice products over patch products.[103,105] The ruvC protein is the only known activity in *E. coli* that demonstrates specificity for such unique structures,[130] arguing that ruvC is the protein charged with the responsibility of resolving of Holliday junctions.

Thus, in separate reactions, both the initiation and resolution phases of homologous recombination have been reconstituted *in vitro*.

IV. Applications of recA protein-mediated homologous DNA targeting

A. Isolation of homologous DNA targets by a recA protein-mediated D-loop

RecA protein can efficiently coat short ssDNA oligonucleotides or ssDNA up to several kilobases in length to form a nucleoprotein filament that is commonly believed to be the active substrate for homologous pairing.[7,9,11,13,16,131] RecA protein-coated ssDNA probes can interact with homologous DNA sequences in a dsDNA target by forming recombination intermediates containing hybridized DNA molecules. These DNA hybrids are believed to exist in unique structures commonly referred to as D-loops or DNA displacement loops.[132,133] The probe-target hybrid molecules are stable when recA protein is removed from D-loops formed with negatively supercoiled DNA targets.[21,134,135] However, deproteinized D-loops are unstable at internal positions in linear dsDNA targets.[43,60,63,131,136-138] The presence of recA protein stabilizes the D-loop in linear dsDNA target molecules.[139]

RecA protein-coated ssDNA can be used to specifically search for and isolate rare sequences in mixtures of homologous and nonhomologous target dsDNA.[41] For example, recA protein-coated circular M13 bacteriophage ssDNA was reacted with a mixture of linear M13 dsDNA and a 1000-fold excess of linearized nonhomologous λ phage DNA. Heterologous "carrier" ssDNA was then added to allow aggregate formation with the nucleoprotein filaments. Next, NaCl was added to a final concentration of 50 m*M* to dissociate the heterologous duplex DNA from the nucleoprotein filament networks. Centrifugation of the reaction allowed purification of the target-homologous DNA hybrid containing the nucleoprotein filaments from the non-sedimenting nonhomologous duplex DNA targets. After 3 additional precipitations, approximately 70% of the homologous duplex target DNA and only 0.04% of the heterologous DNA was recovered. Therefore, the recA protein-mediated targeting provided at least a 1000-fold enrichment procedure for separating molecules containing homologous DNA sequences that were able to efficiently transform competent *E. coli* cells.[41]

DNA as short as a single-stranded 43-mer oligonucleotide could react with negatively supercoiled homologous plasmid DNA targets to form D-loop structures that were stable following removal of the recA protein.[135] This D-loop hybridization reaction using recA protein-coated biotinylated DNA probes was developed into a rapid method for screening supercoiled plasmid libraries.[135] Biotinylated DNA probes coated with recA protein were used to search for homologous sequences in recombinant plasmid libraries consisting of heterogeneous mixtures of negatively supercoiled molecules. Homologous complexes were purified from uncomplexed plasmids by streptavidin-agarose chromatography or with a cupric imidodiacetic acid column. These recovered plasmids were then transformed into *E. coli* and purified. The specific isolation of homologous recombinant plasmid DNA molecules from complex mixtures allowed the recovery of approximately 10 to 20% of the homologous plasmid DNAs and provided a 10,000- to 100,000-fold DNA enrichment of homologous plasmid DNA.[135] This recA protein-mediated D-loop reaction is highly sequence specific; no interaction occurred between the DNA probe and nonhomologous plasmid sequences. Thus, the use of recA protein-mediated D-loop targeting techniques, coupled with affinity selection of the hybrids, permits the isolation of any chosen DNA sequence from complex mixtures of negatively supercoiled recombinant molecules.[135] In contrast to negatively supercoiled DNA targets, neither relaxed circular nor linear duplex DNA targets were present in stable hybrids following removal of recA protein with ssDNA probes targeted to internal positions.

B. Homologous targeting, detection, and purification of linear DNA duplex molecules with two complementary DNA probe strands

A new reaction resulting in formation of a double D-loop structure efficiently targets two relatively short complementary recA protein-coated ssDNA probes to homologous sequences at any position in a linear duplex DNA molecule.[63] Stable hybrids were obtained after recA protein removal when both complementary probe strands were present (4-stranded DNA hybrid), but not when one probe strand was present (3-stranded DNA hybrid). This recA protein-mediated double D-loop targeting reaction provides a new and efficient method to target and detect DNA duplexes, isolate nondenatured duplexes, and map any desired sequence on a duplex DNA or native (nondenatured) chromosomal DNA fragment.[63,139] The native DNA targeting technique has been used for homologous DNA hybridization *in vitro* and in fixed or metabolically active cell nuclei *in situ*.[140]

Double D-loop formation with bacteriophage λ or other linear duplex targets requires the addition of both complementary DNA probe strands for stable hybrid formation; both a Watson and a Crick strand are required for hybrid stability after deproteinization of recA protein. Stable hybrids are formed when the two recA protein-coated complementary probes are added either sequentially or simultaneously.[63] Hybrids formed at internal positions in linear duplex DNA targets with complementary single-stranded DNA probes are stable after recA protein removal, as monitored by co-electrophoresis of ^{32}P-labeled probe and target DNA separated by agarose gel electrophoresis. When only one ssDNA probe strand was used, no hybrid formation was observed.[63]

The specificity of the reaction was demonstrated by hybridizing ^{32}P-radiolabeled recA protein-coated complementary DNA probes to a mixture of restriction endonuclease digested linear duplex fragments of λ DNA. The labeled, recA protein-coated complementary single-stranded DNA probes only hybridized to the expected fragment containing the homologous target region, demonstrating targeting specificity. Hybrid formation was directly proportional to the amount of target fragment added over a 100- to 450-ng range and allowed recA protein to DNA probe nucleotide ratios of 6:1 to 1:3. In addition, the reaction can utilize ATP, ATPγS, or GTPγS as cofactors. The hybrids are stable following SDS, proteinase K and SDS, and phenol extractions.[63]

Biotinylation of one of the complementary single strands allowed the isolation of all the hybrid molecules by affinity capture with streptavidin-coated magnetic beads. Using one strand labeled with ^{32}P and the complementary strand labeled with biotin permits simultaneous detection and capture of native duplex targets.[63] Because both the biotinylated and ^{32}P-labeled strands of the

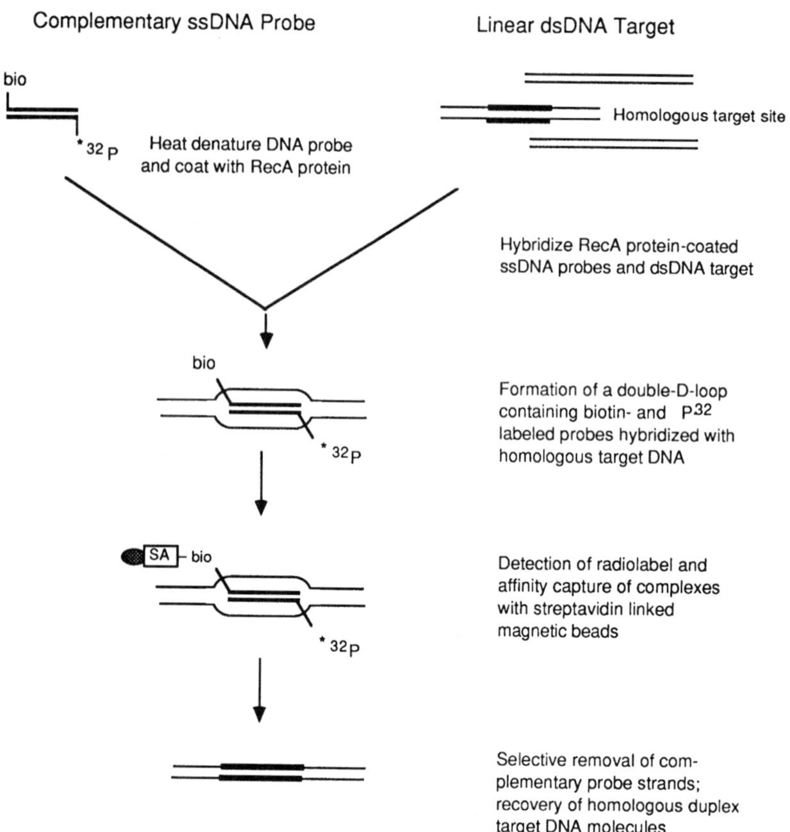

Figure 6. *Targeting, detection, and selective isolation of homologous DNA sequences from heterogeneous mixtures of DNA molecules by recA protein-mediated hybridization of complementary ssDNA probes with target dsDNA. One strand of any desired DNA probe sequence is labeled with biotin (bio), and the complementary DNA strand is radiolabeled with ^{32}P(*). The DNA is heat denatured, coated with recA protein, and hybridized to a complex mixture of homologous and nonhomologous target DNA containing the desired homologous sequence (thick black lines). Hybrids containing both ^{32}P-labeled and biotin-labeled complementary strands of recA protein-coated probe DNA and target DNA were then purified by binding to streptavidin-linked magnetic beads (hatched ellipse labeled "SA-bio"). The amount of ^{32}P-labeled DNA was quantitated by liquid scintillation counting. Duplex target DNA molecules are then recovered intact by selective thermal dissociation of the two probe strands at melting temperatures significantly lower than the duplex target melting temperature.*

probe were present on the probe-target complex, it was proposed that the hybrid molecules contain four DNA strands referred to as a "double-D-loop" structure.[63] Figure 6 shows the simultaneous detection and capture of native (nondenatured) DNA targets. These double-D-loop hybrids were stable at 75°C and most of the probe DNA was removed at 80°C; all the probes dissociated from the duplex after 5 min at 85°C. Because the probe melting temperature is significantly lower than the duplex target melting temperature, the target can be affinity isolated as a probe-target complex and separated from unhybridized probe; this permits recovery of target DNA as a native duplex without ever being heat denatured. The efficiency of deproteinized hybrid recovery generally decreases with complementary DNA probe sizes in the order 500 > 280 > 121 > 79. Approximately 90% of the 500-mer, 65% of the 280-mer, and 5 to 20% of the 121-mer exist in stable double-D-loop hybrids.[63] Therefore, this recA protein-mediated double-D-loop targeting technique allows stabilization of hybrids on linear duplex targets, simultaneous detection and affinity capture of the duplex target, and the recovery of the isolated target DNA as a duplex molecule without any denaturation.

The double-D-loop reaction also allows targeting, identification, and affinity selection of DNA probe sequences homologous with either negatively supercoiled, relaxed, or linear double-stranded DNA target molecules at any position and sequence in the DNA molecules. There, reaction depends on recA protein and a nucleoside triphosphate cofactor, reaction time (60 min), pH (7.5), and temperature (37°C).[63] Double-D-loop formation is sufficiently efficient for targeting complementary ssDNA probes to duplexes identified by only short sequences such as oligonucleotides or sequence-tagged sites (STSs).[141] Furthermore, stable hybrids may be formed with only partial homology, allowing the isolation of homologous DNA containing mutations. Finally, relatively large native chromosomal DNA targets, such as whole 48.5-kb λ phage embedded in agarose, can be recovered as duplex DNA.[63]

Double-D-loop formation in linear DNA duplexes with two complementary recA protein-coated ssDNA strands has been confirmed.[64] The structure of the double-D-loop consists of two duplexes, possibly resembling a DNA replication bubble.[63,64] It was suggested that the double-D-loop may be relevant in recombination reactions involving the combined activities of a helicase and a pairing protein to form intermediates (joint molecules) in homologous recombination,[64] for example, as in the nuclease attenuation model for initiation of homologous recombination by recBCD enzyme.[75,76] The complementary DNA strand that is not involved in the initiation of joint molecule formation in such reactions could, nevertheless, be important in stabilizing D-loop structures by pairing with the displaced strand. In the recA protein and recBCD protein-stimulated pairing reaction (see Figure 5), the intact complementary DNA strand derived from the non-χ-containing strand could stabilize joint molecules as hypothesized.[64]

The double-D-loop technology can be further developed to isolate target sequences from heterogeneous populations and/or sequences with only limited homology to the target, possibly allowing the identification of gene families containing uncharacterized mutations. Large duplex molecules can be isolated intact; thus, there is strong potential for recA protein-mediated hybridization to isolate novel genes from genomic or possibly YAC libraries.[63]

C. RecA protein-assisted restriction endonuclease (RARE) cleavage of chromosomal DNA

The RARE technique may become an important new DNA cloning and mapping method.[137,142] It couples the exquisite homologous targeting specificity of recA protein-mediated D-loop formation to protect any chosen site on a duplex DNA molecule against methylation, allowing subsequent unique cleavage of this desired site. A recA protein-coated single-stranded 30-mer DNA probe can be targeted to a specific λ phage DNA sequence containing an EcoRI restriction site centered in the homologous probe-target site.[137] Incubation of the recA protein-coated synaptic complex with EcoRI methylase and S-adenosylmethionine resulted in the methylation of all the available EcoRI restriction sites, with the important exception of the one protected by the recA protein-coated probe hybridized at the target hybrid site. When recA protein was removed, the oligonucleotide probe dissociated from the target, allowing EcoRI restriction endonuclease cleavage specifically at the homologous site that was protected from methylation, with an overall yield of about 80%.[137]

Specifically chosen DNA fragments were also cleaved from the *E. coli* or human genome using two separate single-stranded recA protein-coated oligonucleotides spanning two different EcoRI sites. These two specific DNA target sequences, located 520 kb apart on the *E. coli* chromosome, were protected by recA protein-mediated hybridization.[137] After methylation and removal of recA protein, cleavage of the 520-kb fragment was observed with a yield of 40% and 60% using 30-mer and 60-mer probes, respectively. Furthermore, a preselected 48- or 180-kb fragment of the human cystic fibrosis transmembrane regulator protein gene (CFTR) was specifically and efficiently cleaved out of human genomic DNA.[137]

RARE cleavage could be expected to occur at approximately every 32 bp given 8 "available" different methyltransferases, each with a different cleavage specificity of 4 base pairs.[142] This could permit precise genomic map distance measurements, as well as the isolation and cloning of any chosen chromosomal DNA fragment. In general, this homologous DNA targeting technique allows recA protein-stabilized single D-loops to specifically and efficiently target any predetermined restriction site for RARE cleavage.[137,142]

D. RecA protein-stabilized D-loop inhibits transcription of DNA by RNA polymerase

RecA protein-coated ssDNA also can be targeted to homologous internal promoter sequences in linear dsDNA to form protein-stabilized D-loops. Such D-loops were positioned in a promoter region to block the transcription of a linear dsDNA target by RNA polymerase.

RecA protein-coated joint molecule complexes covering all or a portion of the T7 or T3 phage promoter blocked RNA transcription by T7 or T3 RNA polymerases, respectively.[143] The inhibition of T7 transcription with a recA protein-coated ssDNA 42-mer homologously targeted to either DNA strand inhibited RNA transcription by more than 98% in each case. No inhibition of transcription occurred with recA protein-coated nonhomologous DNA or with a recA protein-stabilized D-loop formed 93 bases upstream of the T3 promoter, indicating that the D-loop may provide a direct barrier to the polymerase. A homologous recA protein-coated 70-mer DNA covering the promoter inhibited transcription by >99%, and a 20-mer covering either 12 or 17 bases of the T3 promoter inhibited transcription by about 93%.

The elongation of transcription was also inhibited when the recA protein-stabilized D-loop was positioned 276 to 317 bp downstream from the T7 promoter.[143] The formation of this D-loop resulted in a truncated RNA transcript of the expected size with both T3 and T7 RNA polymerases. Furthermore, in addition to truncated DNA transcripts, the RNA polymerase molecules may be physically retained by the recA protein-stabilized D-loop because the D-loop-arrested RNA polymerase molecules appear not to be involved in any further transcriptional events.

These studies were extended to test for transcriptional inhibition of human RNA polymerase II by recA protein-stabilized D-loops.[144] RecA protein-coated D-loops formed with ssDNA homologous to a viral TATA box (which interacts with cellular factor TFIID) or to three GC boxes (which bind cellular factor Sp1) inhibited transcription by RNA polymerase II. Complete inhibition of the eukaryotic RNA polymerase was observed with oligonucleotides 33 nucleotides in length, but was less complete with 22 nucleotides, whereas 20 nucleotides were sufficient with the phage RNA polymerase.[143,144] Experiments with T3 and T7 promoters showed D-loops targeted downstream of the RNA polymerase promoter inhibited transcription elongation and produced truncated transcripts;[143] however, D-loops formed downstream of the HIV-1 promoter did not block elongation of RNA[144] by RNA polymerase II. The stability of this recA protein-stabilized D-loop was demonstrated by specific inhibition of restriction endonuclease cleavage at the DraI restriction endonuclease site.[144] Thus, recA protein-hybridized oligonucleotides can be used to block both eukaryotic and prokaryotic RNA polymerases, as well as prokaryotic methylases or restriction endonucleases.

It was suggested that the recA protein-stabilized single D-loop may be used as an alternative method to the mutational analysis of promoters of transcription.[144] It will also be interesting to determine whether recA protein-stabilized D-loop hybrids can also be used to probe DNA sites that interact with different sequence-specific DNA proteins other than those involved in RNA transcription. It is not known whether the use of recA protein-stabilized D-loops may provide a general method for targeting homologous single-stranded oligonucleotides to any sequence to inhibit other important DNA enzymes, for example, DNA polymerase.

E. RecA protein-mediated targeting and mapping of homologous ssDNA probes to sites on individual DNA molecules analyzed by high-resolution darkfield electron microscopy

RecA protein-coated DNA probes have been successfully used for directly visualizing and mapping the sites of homologous probe-target hybrids on individual linear dsDNA molecules by high-resolution darkfield electron microscopy (EM). Linear DNA was reacted with recA protein protein-coated complementary ssDNA probes that were 446 or 222 bases in length to measure the accuracy of the targeting reaction as evidenced by the position of the recA protein-coated DNA probe on individual target molecules.[139] This is possible because the recA protein-coated ssDNA nucleoprotein filament is approximately five times the diameter of uncoated dsDNA target molecules. In reactions using recA protein-coated complementary pairs of ssDNA, the protein-coated target sites appear as distinct thick products on the DNA strand.

Figure 7 shows the progression of the targeting reaction at 37°C and the mapping of a 446-base-long probe on individual linear duplex molecules analyzed by darkfield EM.[139] Figure 7, Panel A, shows the short duplex 446-mer DNA probe. Following heat denaturation and rapid cooling, single strands of the 446-mer are separated and appeared as bushes or filaments in the background (Panel B). After 5 to 15 min, the ssDNA probes covered with recA protein were clearly visible and the nucleoprotein filaments were more rigid and thicker than uncoated DNA[16,145,146] (Figure 7, Panel C). The arrow points to a minor population of DNA probe molecules that probably renatured before they could be coated with recA protein. Coating is complete after 15 min at 37°C, as judged by darkfield EM. Approximately 20 min after incubation of target duplex DNA with the recA protein-coated 446-mer, large DNA synaptic complexes are formed. These web-like synaptic complexes are composed of molecules with DNA probe homologies located at either end of the target molecules (Panel D). Panel E shows the 446-mer hybridized to homologous sites at either end of the target after 45 to 60 min at 37°C. At higher resolution, a target molecule is shown with the probe at a homologous end site. The arrow shows the position of a physically bent DNA marker near the other end of this target (Panel F). Panel G shows the 446-mer after 15 to 20 min reaction positioned to a homologous site in another target near the bent DNA marker. Finally, Figure 7, Panel H, shows the 446-mer probe precisely paired with the homologous target site adjacent to the bent DNA marker.

In general, hybridization is complete in 45 min and all the DNA probes are bound. No hybrids are formed with nonhomologous DNA targets. Thus, darkfield EM mapping of recA protein-coated ssDNA probes on individual linear duplex molecules allows direct visualization of individual hybrids at high resolution without chemical fixation or sample shadowing.[139] Therefore, there is no DNA distortion or shrinkage. The hybridization reaction is highly efficient and

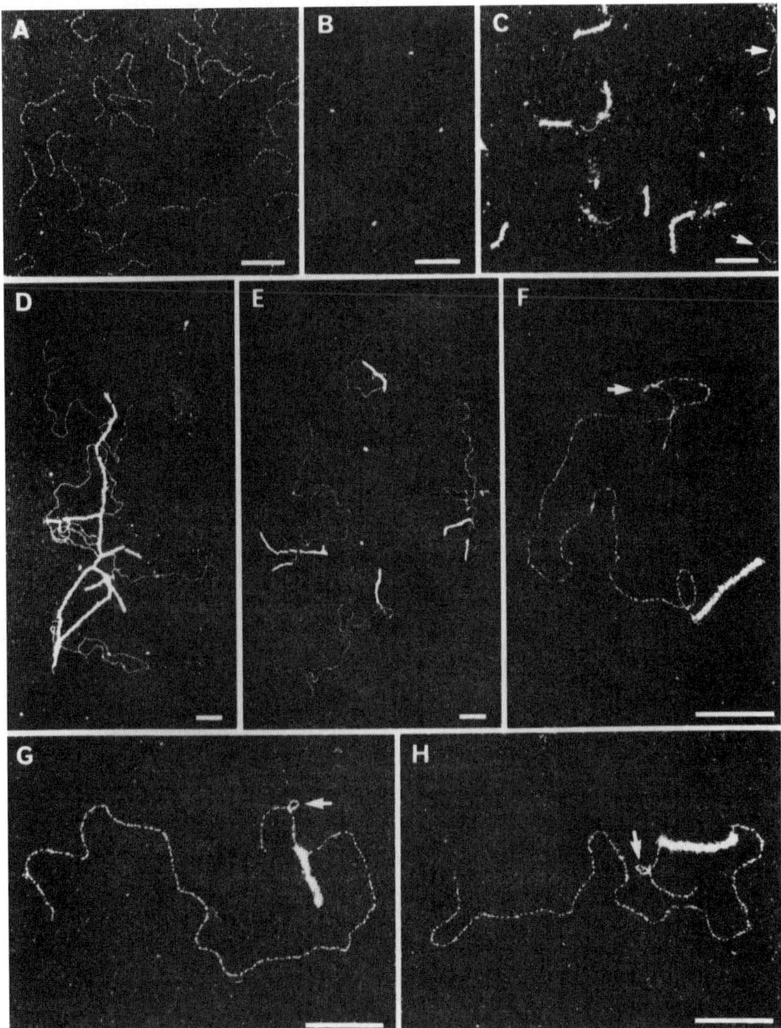

Figure 7. *Homologous targeting on linear duplex molecules with complementary ssDNA probes coated with recA protein and analysis by annular darkfield electron microscopy. A 446-base-long ssDNA probe was heat denatured, coated with recA protein, and reacted with a linear duplex plasmid DNA target. The DNA target ends were defined by restriction endonuclease cleavage sites and contain a physically bent DNA marker for mapping the DNA probe position to the homologous target site. Panels* ***A—H*** *are described in the text. The lengths of the bars represent 0.1 mm. (From Révet, B.M., Sena, E.P., and Zarling, D.A.,* J. Mol. Biol., *232, 779, 1993. With permission.)*

mapping of probes is extremely precise, to within 40 nucleotides, as judged by high-resolution darkfield EM. This targeting and mapping technique should allow the use of short probes, including STSs, to visually map sequences on cloned DNA and large chromosomal DNA fragments using either electron or laser microscopic analyses.

Homologous targeting was also monitored at two different sites separated by 900 bp in the DNA target.[139] Two different 446- and 222-base-long probes can both specifically target their homologous target sequences on the same molecule, but not as frequently as expected. When added at equimolar concentrations, the 446- and 222-base-long probes were both observed at each respective site (separated by the expected 900 bases) on the same molecule in only 2% of the total DNA molecules. About 40% of the targets hybridized the 446-mer alone, and 40% of the targets hybridized the 222-mer alone, but only 2% of the targets had both the 446-mer and the 222-mer probes positioned together (900 bp apart) on the same molecule. It was hypothesized that topological strains may be imposed in the target molecules by the directionality of probe movement in the homology search and/or separation of DNA strands in the target region.[139]

In summary, the DNA probe-target hybrids can be visualized on individual DNA molecules at any sequence, and the recA protein serves as a marker for the probe's position on the target. These recA protein-mediated targeting reactions and direct visual mapping techniques should allow rapid and precise DNA sequence mapping on individual DNA molecules, on large chromosomal DNA fragments, or even on whole chromosomes.

F. RecA protein-mediated targeting of cellular and viral chromosomal DNA in fixed and metabolically active nuclei analyzed by native FISH

RecA protein-mediated fluorescence *in situ* hybridization (FISH) methods could allow homologous DNA targeting to nondenatured chromosomal DNA targets. A new and powerful native FISH method has been developed for targeting of complementary ssDNA probes coated with recA protein to homologous nondenatured human viral and cellular chromosomal DNA sequences.[140,147]

In classic FISH reactions, methanol:acetic fixed and permeabilized cellular DNA is denatured and hybridized with ssDNA probes in mass action driven DNA-DNA hybridization reactions.[148] The kinetics of the hybridization reaction is determined by the DNA probe and target DNA concentrations, reaction time, ionic concentration of the buffer, and temperature. Hybridization requires thermal denaturation of the chromosomal target DNA.[148]

Native FISH methods eliminate the requirement for denaturing the duplex DNA to produce single-stranded target sequences for hybridization. Classic FISH methods treat target DNA in fixed cells by either heat denaturation, chemical denaturation, or endonuclease digestion to prepare the DNA target for hybridization. These severe denaturation treatments may produce spurious and unwanted changes in metaphase or interphase DNA structure and repeated sequence random reassociations, and the lengthy hybridization reaction times add significant time to the FISH procedure.

A sequence-specific, rapid, and efficient recA protein-mediated native-DNA FISH reaction has been developed that is fully capable of detecting either multiple-copy or single-copy gene sequences in fixed cells probed on slides or in suspension.[140,147] The recA protein-mediated native FISH reactions were observed with chromosome-specific satellite centromeric DNA probes specific for human chromosome X and 7 DNA sequences in HEp-2 human epitheloid carcinoma cells, and chromosomes 1 and 17 in human hepatitis B virus (HBV)-infected hepatocellular carcinoma cells (HCC). Native chromosome 17 p53 tumor suppressor genes in HEp-2 or HCC cells were detected with p53 cDNA as well as with human chromosome 17 alpha satellite DNA probes. Also, HBV DNA targets were observed in HCC cells. These DNA target sequences were visualized with 100- to 600-base-long or larger biotinylated (nick-translated with bio-14-dUTP) probes by FISH and analyzed by fluorescent microscopy or confocal laser scaling microscopy (CLSM). In these reactions, there was no amplification of the FITC-labeled biotinylated DNA probe.[148]

Figure 8, top, shows nondenatured acetic acid:ethanol fixed HEp-2 cells hybridized with human chromosome 7 specific alpha satellite centromeric DNA probe coated with recA protein. Under these conditions most of the interphase HEp-2 nuclei cell on slides showed about 2 FITC-labeled FISH signals. No FISH signals were obtained in reactions without recA protein or without incubation at 37°C. Comparisons of the number of chromosome FISH signals per HEp-2 cell nucleus were similar using the native FISH reaction compared with the control denatured FISH reaction (Figure 8, bottom). The FITC signals from similar native FISH reactions with human chromosome 1 were located at the centromere region of human chromosome 1 in metaphase spreads from HEp-2 nuclei. HEp-2 cells have 2 to 3 number 1 chromosomes and show approximately 2 to 3 FISH signals in both interphase nuclei and metaphase chromosome spreads. In these experiments, about 73% of the interphase cell nuclei showed hybridization signals.[147]

In general, alpha satellite probes in the range of about 100 to 600 bases in length produced recA protein-mediated native FISH signals. In different experiments, the strongest FISH signals were usually obtained with probes averaging 400 bases in length produced by nick translation of the DNA with bio-14-dUTP. The native recA protein-mediated FISH reaction with the human chromosomal 1 probe was proportional to the probe concentration over the range of 10 to 50 ng/reaction. The kinetics of the recA protein-mediated FISH reaction showing hybridization with fixed nuclei in solution was maximal at approximately 2 hr, and on slides was optimal at about 2.5 hr, following probe addition. Native FISH reactions required an ATPγS or ATP cofactor and recA protein/DNA probe nucleotide ratios of 1:1 or 1:2.

This native *in situ* hybridization technique with fixed cells was adapted for native DNA targeting in metabolically active nuclei.[140,147] Efficient dsDNA

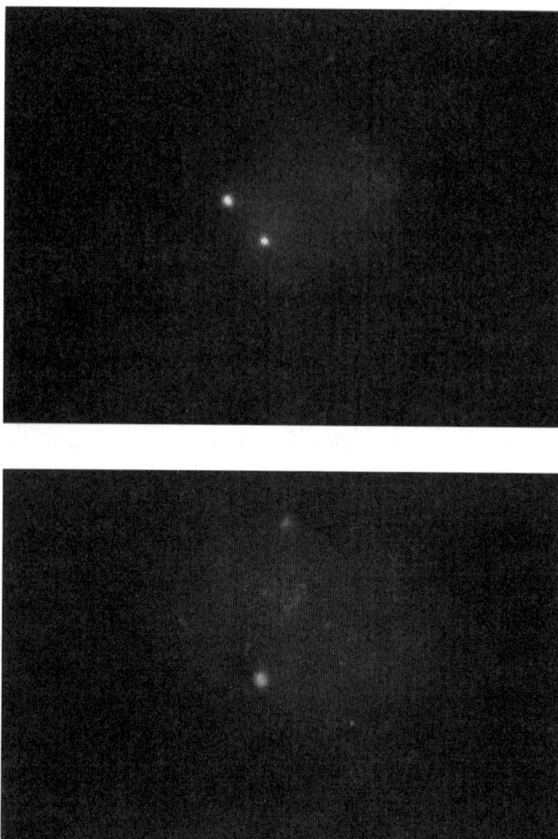

*Figure 8. Comparison of recA protein-mediated fluorescence in situ hybridization of human chromosome 7 alpha satellite probe DNA with nondenatured HEp-2 target cell nuclei (**top**), with classic FISH (not recA protein-mediated) in heat denatured target cell nuclei (**bottom**). HEp-2 interphase nuclei were fixed with acetic acid and methanol (3:1) and probed on slides with biotinylated (nick-translated) ssDNA probes specific for human chromosome number 7 alpha satellite DNA sequences. In the reaction shown in the top of Figure 8 the recA protein coated complementary single-stranded chromosome 7 DNA probes were reacted with nondenatured nuclei for 2.5 hr at 37°C, followed by washing, blocking, and adding FITC-avidin as the reporter. In the reaction shown in the bottom, heat-denatured nuclei were hybridized overnight with heat-denatured chromosome 7 alpha satellite DNA, washed, and blocked, and FITC-avidin was also used as the reporter. In the denatured FISH reaction, the signal was amplified by the addition of a biotinylated goat anti-avidin antibody.*

targeting techniques were developed in cell nuclei embedded in agarose. Human cell nuclei permeabilized with the non-ionic detergent Triton-100 and encapsulated in agarose remain morphologically intact, synthesize both DNA and RNA, and permit entry of recA protein-coated ssDNA probes.[140] FISH in metabolically active nuclei was tested using recA protein-coated ssDNA complementary 200- to 800-base-long biotinylated repeated satellite or unique sequence DNA probes with a streptavidin-FITC reporter and viewed by fluorescence microscopy or CLSM.[140,147]

Figure 9 shows p53 DNA probes targeted in metabolically active human HEp-2 cell nuclei embedded in agarose. In these targeting reactions, 60 ng of recA protein-coated single-stranded complementary biotinylated (nick-translated) p53 cDNA was reacted for 3 hr at 37°C. The two FITC-labeled nuclear FISH signals are overlaid onto the phase image of the nuclei in the top left panel (Figure 9).

The native DNA targeting and hybridization reaction with biotinylated (nick-translated) p53 cDNA probes in nuclei embedded in agarose required

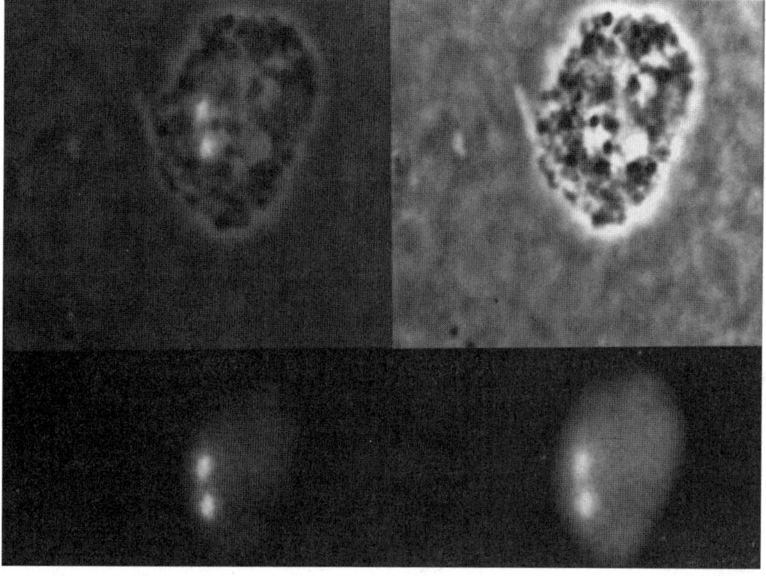

Figure 9. *Targeting of recA protein-coated complementary single-stranded p53 DNA to native (nondenatured) chromosomal DNA sequences in metabolically active human cell nuclei. Nuclei of HEp-2 cells encapsulated in agarose[151] were reacted with recA protein-coated biotinylated p53 cDNA probes for 3 hr at 37°C and washed. FITC-avidin was added as a reporter and the nuclei were washed and immediately visualized by confocal laser scanning microscopy. The phase image of a metabolically active HEp-2 cell nuclei in agarose is shown in the upper right panel. The lower right panel shows a digital image of a nucleus with two p53 DNA signals as monitored by laser scanning microscopy. This digital image was processed and the processed image is shown in the lower left panel. The processed digital image was overlaid onto the phase image in the top left panel.*

incubation with recA protein at 37°C, was directly proportional to DNA probe concentration in the range of 30 to 90 ng of DNA probe per reaction, pH (7.4), and required a nucleoside triphosphate cofactor (ATP or ATPγS). Targeting required a 3-h incubation at 37°C of recA protein-coated complementary ssDNA DNA probes with the permeabilized nuclei. Similar numbers of native FISH signals were observed in metabolically active nuclei as in interphase fixed cells permeabilized with acetic acid and methanol using either recA protein-mediated nondenatured DNA FISH, classic heat-denatured FISH in fixed interphase nuclei, or metaphase chromosomes using either unique or repeated sequence-specific DNA probes. When human satellite DNA probes coated with recA protein were hybridized with metabolically active nuclei and treated with hypotonic buffers to decondense the chromatin, repeated chromosomal specific-satellite FISH signals were specifically localized along the chromatin fiber. In intact nuclei-repeated satellite or unique p53 gene, FISH signals are often paired following nuclear DNA replication, as observed by confocal laser scanning microscopy and digital image analysis.[140,147]

In summary, the recA protein-mediated *in situ* targeting reaction uses complementary nick-translated DNA probes with native (nondenatured) target sequences. The hybridization reaction depends on recA protein cofactor (ATP or ATPγS, 0.48 m*M*) and reaction time (2 to 2.5 hr), pH (7.5), and temperature (37°C). Nick-translated cDNA probes efficiently form hybrids in about 75 to 85% of the nuclei. The recA protein-mediated targeting to DNA in fixed cells can be detected without signal amplification. Similar numbers of nondenatured FISH signals compared with heat-denatured FISH signals are observed. The specificity of the reaction was evidenced by homologous competition, copy number, linkage to well-established chromosomal markers, and metaphase cytogenetics.[147]

G. Potential applications of RecA protein-mediated homologous DNA targeting

These recA protein-mediated homologous DNA targeting reactions may be applied to target, detect, isolate, clone, and map any desired duplex gene or homologous chromosomal DNA fragment.[41,63,64,135,139] Novel gene mapping methods are evolving either by microscopy of large molecules or whole chromosomes, which may be physically oriented and stretched. RecA protein-mediated native chromosome walking techniques would permit more rapid gene discovery and mapping procedures. Furthermore, targeting of ssDNA probes to specific homologous sites by recA protein might block transcription *in vivo* as well as *in vitro*[143,144] DNA methylation[137,142] or DNA cleavage.[143] RecA protein-mediated targeting has been used to detect chromosomal targets in both fixed and metabolically active nuclei. This important new *in situ* DNA targeting technique could possibly be also used for gene tracking studies. Lastly, it may be possible to use recA protein-mediated DNA targeting and homologous DNA transfer for recombination enzyme-mediated gene therapy as hypothesized in Figure 10.

V. Conclusions

Both the molecular mechanisms of homologous recombination and the biochemical behavior of recombination enzymes are becoming increasingly clear. With increased knowledge, it should be possible to design more effective *in vivo* gene targeting strategies in biological systems that have been recalcitrant to efficient and useful homologous recombination protocols. Until it becomes clear why many integration events in mammalian cells are nonhomologous, specific suggestions for improvement are not conspicuous; however, several general ideas, based on the paradigms established in *E. coli*, present themselves. First, rather than simply introducing foreign target DNA

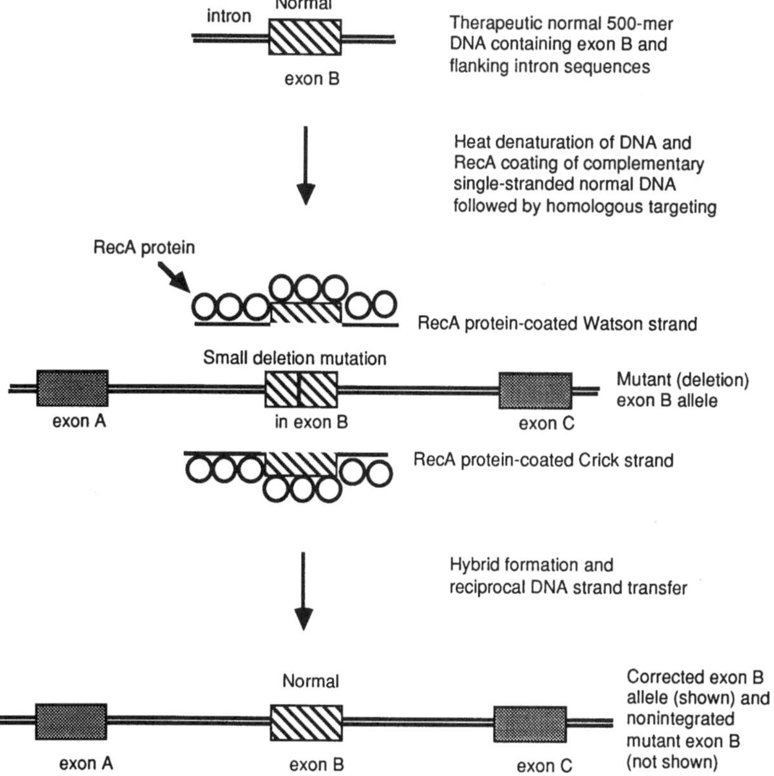

Figure 10. *Possible mechanism for homologous DNA targeting and transfer for gene correction of a mutant allele using recA protein coated relatively short (e.g., 500-mer) complementary ssDNA probes.*

into cells, it might be advantageous to co-introduce enzymes that enhance various steps of the recombination mechanism. If, for example, dsDNA break sites are limiting, then an endonuclease can be introduced. Similarly, if generation of ssDNA from the donor DNA is limiting, then either ssDNA itself, or helicases or nucleases to maintain the single-stranded form can be in introduced or co-expressed. Recombination hotspots can be introduced into the donor DNA, or alternatively, mutant or modified recombination proteins with constitutively activated or enhanced recombination functions can be used. Alternatively, the over-expression of normally limiting endogenous cellular recombination enzymes, in concert with other normal cellular components, may be harnessed to increase recombination efficiency.

The applications of recA protein-targeted events present themselves as immediate options (for example, see Figure 6). Interestingly, when a modified *E. coli* recA protein containing a nuclear localization signal is expressed in mammalian cells, it localizes to the nucleus and binds to chromosomes.[149] Perhaps, similar expression vectors can be used to introduce other recombination enzymes into the nucleus of mammalian cells. The continued taming and manipulation of the homologous recombination process can afford a variety of new options for more effective gene targeting protocols.

Acknowledgments

The authors are indebted to Angela Eggleston for providing Figure 3 and Elissa Sena and Sushma Patel for helpful discussions. The work in the authors' laboratories is supported by grants from the National Institutes of Health (AI-18987 and GM-41347 to S.C.K. and GM-38424 to D.A.Z.).

References

1. Gottesman, M.E., and Vogel, H.J., Ed., *Mechanisms of Eukaryotic DNA Recombination*, Academic Press, Inc., San Diego, 1992.

2. Sedivy, J.M., and Joyner, A.L., *Gene Targeting*, W.H. Freeman and Company, New York, 1992, 184.

3. Lin, M., Sperle, K., and Sternberg, N., Studies on extrachromosomal homologous recombination in mammalian cells: implications for chromosomal recombination and gene targeting, in *Mechanisms of Eukaryotic DNA Recombination*, Gottesman, M.E., and Vogel, H.J., Eds., Academic Press, Inc., San Diego, 1992, 15.

4. Cox, M.M., and Lehman, I.R., Enzymes of general recombination, *Annu. Rev. Biochem.*, 56, 229, 1987.

5. Eggleston, A.K., and Kowalczykowski, S.C., An overview of homologous pairing and DNA strand exchange proteins, *Biochimie*, 73, 163, 1991.

6. Fishel, R., General genetic recombination, in *Modern Microbial Genetics*, Wiley-Liss, New York, 1991, 91.

7. Griffith, J.D., and Harris, L.D., DNA strand exchanges, *CRC Crit. Rev. Biochem.*, 23 Suppl. 1, S43, 1988.

8. Heyer, W.-D., and Kolodner, R.D., Enzymology of homologous recombination in *Saccharomyces cerevisiae*, *Prog. Nucleic Acid Res. Mol. Biol.*, 46, 221, 1993.

9. Kowalczykowski, S.C., Biochemistry of genetic recombination: energetics and mechanism of DNA strand exchange, *Annu. Rev. Biophys. Biophys. Chem.*, 20, 539, 1991.

10. Kowalczykowski, S.C., Dixon, D.A., Eggleston, A.K., Lauder, S. D., and Rehrauer, W.M., Biochemistry of homologous recombination in *Escherichia coli*, *Microbiol. Rev.*, 58, 3, 1994.

11. Kowalczykowski, S.C., and Eggleston, A.K., Homologous pairing and DNA strand exchange proteins, *Annu. Rev. Biochem.*, 63, 991, 1994.

12. Lloyd, R.G., and Sharples, G.J., Genetic analysis of recombination in prokaryotes, *Curr. Opin. Genet. Dev.*, 2, 683, 1992.

13. Radding, C.M., Helical RecA nucleoprotein filaments mediate homologous pairing and strand exchange, *Biochim. Biophys. Acta*, 1008, 131, 1989.

14. Radding, C.M., Helical interactions in homologous pairing and strand exchange driven by RecA protein, *J. Biol. Chem.*, 266, 5355, 1991.

15. Roca, A.I., and Cox, M.M., The RecA protein: structure and function, *Crit. Rev. Biochem. Mol. Biol.*, 25, 415, 1990.

16. Stasiak, A., and Egelman, E.H., Visualization of recombination reactions, in *Genetic Recombination*, Kucherlapati, R., and Smith, G.R., Eds., American Society for Microbiology, Washington, DC, 1988, 265.

17. West, S.C., Enzymes and molecular mechanisms of genetic recombination, *Annu. Rev. Biochem.*, 61, 603, 1992.

18. Thaler, D.S., and Stahl, F.W., DNA double-chain breaks in recombination of phage lambda and of yeast, *Annu. Rev. Genet.*, 22, 169, 1988.

19. Smith, G.R., Homologous recombination in procaryotes, *Microbiol. Rev.*, 52, 1, 1988.

20. Smith, G.R., Conjugational recombination in *E. coli*: myths and mechanisms, *Cell*, 64, 19, 1991.

21. Konforti, B.B., and Davis, R.W., 3′ homologous free ends are required for stable joint molecule formation by the RecA and single-stranded binding proteins of *Escherichia coli.*, *Proc. Natl. Acad. Sci. U.S.A.*, 84, 690, 1987.

22. West, S.C., and Connolly, B., Biological roles of the *Escherichia coli* RuvA, RuvB and RuvC proteins revealed, *Mol. Microbiol.*, 6, 2755, 1992.

23. Shinohara, A., Ogawa, H., and Ogawa, T., Rad51 protein involved in repair and recombination in *S. cerevisiae* is a RecA-like protein, *Cell*, 69, 457, 1992.

24. Shinohara, A., Ogawa, H., Matsuda, Y., Ushio, N., Ikeo, K., and Ogawa, T., Cloning of human, mouse and fission yeast recombination genes homologous to *RAD51* and *recA*, *Nat. Genet.*, 4, 239, 1993.

25. Cox, M.M., and Lehman, I.R., RecA protein of *Escherichia coli* promotes branch migration, a kinetically distinct phase of DNA strand exchange, *Proc. Natl. Acad. Sci. U.S.A.*, 78, 3433, 1981.

26. Kowalczykowski, S.C., Clow, J., and Krupp, R.A., Properties of the duplex DNA-dependent ATPase activity of *Escherichia coli* recA protein and its role in branch migration, *Proc. Natl. Acad. Sci. U.S.A.*, 84, 3127, 1987.

27. Pugh, B.F., and Cox, M.M., Stable binding of recA protein to duplex DNA. Unraveling a paradox, *J. Biol. Chem.*, 262, 1326, 1987.

28. Pugh, B.F., and Cox, M.M., General mechanism for RecA protein binding to duplex DNA, *J. Mol. Biol.*, 203, 479, 1988.

29. West, S.C., Cassuto, E., Mursalim, J., and Howard-Flanders, P., Recognition of duplex DNA containing single-stranded regions by recA protein, *Proc. Natl. Acad. Sci. U.S.A.*, 77, 2569, 1980.

30. Shaner, S.L., and Radding, C.M., Translocation of *Escherichia coli* recA protein from a single-stranded tail to contiguous duplex DNA, *J. Biol. Chem.*, 262, 9211, 1987.

31. Blaho, J.A., and Wells, R.D., Left-handed Z-DNA binding by the recA protein of *Escherichia coli*, *J. Biol. Chem.*, 262, 6082, 1987.

32. Kim, J.I., Heuser, J., and Cox, M.M., Enhanced recA protein binding to Z DNA represents a kinetic perturbation of a general duplex DNA binding pathway, *J. Biol. Chem.*, 264, 21848, 1989.

33. Thresher, R.J., and Griffith, J.D., Intercalators promote the binding of RecA protein to double-stranded DNA, *Proc. Natl. Acad. Sci. U.S.A.*, 87, 5056, 1990.

34. Kojima, M., Suzuki, M., Morita, T., Ogawa, T., Ogawa, H., and Tada, M., Interaction of RecA protein with pBR322 DNA modified by N-hydroxy-2-acetylaminofluorene and 4-hydroxyaminoquinoline 1-oxide, *Nucleic Acids Res.*, 18, 2707, 1990.

35. Dianov, G.L., Saparbaev, M.K., Mazin, A.V., and Salganik, R.I., The chemical mutagen dimethyl sulphate induces homologous recombination of plasmid DNA by increasing the binding of RecA protein to duplex DNA, *Mutat. Res.*, 249, 189, 1991.

36. Stasiak, A., DiCapua, E., and Koller, T.H., Elongation of duplex DNA by recA proteins, *J. Mol. Biol.*, 151, 557, 1981.

37. Menetski, J.P., and Kowalczykowski, S.C., Interaction of recA protein with single-stranded DNA. Quantitative aspects of binding affinity modulation by nucleotide cofactors, *J. Mol. Biol.*, 181, 281, 1985.

38. Conley, E.C., and West, S.C., Underwinding of DNA associated with duplex-duplex pairing by RecA protein, *J. Biol. Chem.*, 265, 10156, 1990.

39. Lindsley, J.E., and Cox, M.M., On RecA protein-mediated homologous alignment of two DNA molecules. Three strands versus four strands, *J. Biol. Chem.*, 265, 10164, 1990.

40. Hsieh, P., Camerini-Otero, C.S., and Camerini-Otero, R.D., The synapsis event in the homologous pairing of DNAs: RecA recognizes and pairs less than one helical repeat of DNA, *Proc. Natl. Acad. Sci. U.S.A.*, 89, 6492, 1992.

41. Honigberg, S.M., Rao, B.J., and Radding, C.M., Ability of RecA protein to promote a search for rare sequences in duplex DNA, *Proc. Natl. Acad. Sci. U.S.A.*, 83, 9586, 1986.

42. Honigberg, S.M., Gonda, D.K., Flory, J., and Radding, C.M., The pairing activity of stable nucleoprotein filaments made from recA protein single-stranded DNA and adenosine 5′-γ-thiotriphosphate, *J. Biol. Chem.*, 260, 11845, 1985.

43. Riddles, P.W., and Lehman, I.R., The formation of plectonemic joints by the recA protein of *Escherichia coli*, *J. Biol. Chem.*, 260, 170, 1985.

44. Menetski, J.P., Bear, D.G., and Kowalczykowski, S.C., Stable DNA heteroduplex formation catalyzed by the *Escherichia coli* RecA protein in the absence of ATP hydrolysis, *Proc. Natl. Acad. Sci. U.S.A.*, 87, 21, 1990.

45. Rosselli, W., and Stasiak, A., Energetics of RecA-mediated recombination reactions: without ATP hydrolysis RecA can mediate polar strand exchange but is unable to recycle, *J. Mol. Biol.*, 216, 335, 1990.

46. Kim, J.I., Cox, M.M., and Inman, R.B., On the role of ATP hydrolysis in RecA protein-mediated DNA strand exchange. I. Bypassing a short heterologous insert in one DNA substrate, *J. Biol. Chem.*, 267, 16438, 1992.

47. Kowalczykowski, S.C., and Krupp, R.A., DNA strand exchange promoted by *Escherichia coli* recA protein in the absence of ATP: implications for the mechanism of energy transduction, submitted, 1994.

48. Gonda, D.K., and Radding, C.M., The mechanism of the search for homology promoted by recA protein: facilitated diffusion within nucleoprotein networks, *J. Biol. Chem.*, 261, 13087, 1986.

49. West, S.C., Cassuto, E., and Howard-Flanders, P., Homologous pairing can occur before DNA strand separation in general genetic recombination, *Nature (London)*, 290, 29, 1981.

50. Bianchi, M., Riboli, B., and Magni, G., *Escherichia coli* recA protein possesses a strand separating actvity on short duplex DNA, *EMBO J.*, 4, 3025, 1985.

51. Hsieh, P., and Camerini-Otero, R.D., Formation of joint DNA molecules by two eukaryotic strand exchange proteins does not require melting of a DNA duplex, *J. Biol. Chem.*, 264, 5089, 1989.

52. Howard-Flanders, P., West, S.C., and Stasiak, A., The spiral protein filaments of genetic recombination, *Nature (London)*, 309, 1, 1983.

53. Stasiak, A., Three-stranded DNA structure; is this the secret of DNA homologous recognition?, *Mol. Microbiol.*, 6, 3267, 1992.

54. Camerini-Otero, R.D., and Hsieh, P., Parallel DNA triplexes, homologous recombination, and other homology-dependent DNA interactions, *Cell*, 73, 217, 1993.

55. Stasiak, A., DiCapua, E., and Koller, T., Unwinding of duplex DNA in complexes with recA protein, *Cold Spring Harbor Symp. Quant. Biol.*, 47, 811, 1983.

56. Wu, A.M., Bianchi, M., DasGupta, C., and Radding, C.M., Unwinding associated with synapsis of DNA molecules by recA protein, *Proc. Natl. Acad. Sci. U.S.A.*, 80, 1256, 1983.

57. Schutte, B.C., and Cox, M.M., Homology-dependent underwinding of duplex DNA in recA protein generated paranemic complexes, *Biochemistry*, 27, 7886, 1988.

58. Honigberg, S.M., and Radding, C.M., The mechanics of winding and unwinding helices in recombination: torsional stress associated with strand transfer promoted by RecA protein., *Cell*, 54, 525, 1988.

59. Rould, E., Muniyappa, K., and Radding, C.M., Unwinding of heterologous DNA by RecA protein during the search for homologous sequences, *J. Mol. Biol.*, 226, 127, 1992.

60. Riddles, P.W., and Lehman, I.R., The formation of paranemic and plectonemic joints between DNA molecules by the recA and single-stranded DNA-binding proteins of *Escherichia coli*, *J. Biol. Chem.*, 260, 165, 1985.

61. Cunningham, R.P., Wu, A.M., Shibata, T., DasGupta, C., and Radding, C.M., Homologous pairing and topological linkage of DNA molecules by combined action of *E. coli* recA protein and topoisomerase I, *Cell*, 24, 213, 1981.

62. Cassuto, E., Formation of covalently closed heteroduplex DNA by the combined action of gyrase and RecA protein, *EMBO J.*, 3, 2159, 1984.

63. Sena, E.P., and Zarling, D.A., Targeting in linear DNA duplexes with two complementary probe strands for hybrid stability, *Nat. Genet.,* 3, 365, 1993.

64. Jayasena, V.K., and Johnston, B.H., Complement-stabilized D-loop. RecA-catalyzed stable pairing of linear DNA molecules at internal sites, *J. Mol. Biol.*, 230, 1015, 1993.

65. Bianchi, M.E., and Radding, C.M., Insertions, deletions and mismatches in heteroduplex DNA made by recA protein, *Cell*, 35, 511, 1983.

66. Ramdas, J., Mythili, E., and Muniyappa, K., RecA protein promoted homologous pairing *in vitro.* Pairing between linear duplex DNA bound to HU protein (nucleosome cores) and nucleoprotein filaments of recA protein-single-stranded DNA, *J. Biol. Chem.*, 264, 17395, 1989.

67. Ramdas, J., Mythili, E., and Muniyappa, K., Nucleosomes on linear duplex DNA allow homologous pairing but prevent strand exchange promoted by recA protein, *Proc. Natl. Acad. Sci. U.S.A.*, 88, 1344, 1991.

68. Muniyappa, K., Ramdas, J., Mythili, E., and Galande, S., Homologous pairing between nucleosome cores on a linear duplex DNA and nucleoprotein filaments of RecA protein-single stranded DNA, *Biochimie*, 73, 187, 1991.

69. Clark, A.J., Sharma, V., Brenowitz, S., Chu, C.C., Sandler, S., Satin, L., Templin, A., Berger, I., and Cohen, A., Genetic and molecular analyses of the C-terminal region of the *recE* gene from the rac prophage of *Escherichia coli* K-12 reveal the *recT* gene, *J. Bacteriol.*, 175, 7673, 1993.

70. Hall, S.D., Kane, M.F., and Kolodner, R.D., Identification and characterization of the *Escherichia coli* RecT protein, a protein encoded by the *recE* region that promotes renaturation of homologous single-stranded DNA, *J. Bacteriol.*, 175, 277, 1993.

71. Hall, S.D., and Kolodner, R.D., Homologous pairing and strand exchange promoted by the *Escherichia coli* RecT protein, *Proc. Natl. Acad. Sci. U.S.A.*, 91, 3205, 1994.

72. Sander, M., Carter, M., and Huang, S.M., Expression of Drosophila Rrp1 protein in *Escherichia coli.* Enzymatic and physical characterization of the intact protein and a carboxyl-terminally deleted exonuclease-deficient mutant, *J. Biol. Chem.*, 268, 2075, 1993.

73. Taylor, A.F., The recBCD enzyme of *Escherichia coli*, in *Genetic Recombination*, Kucherlapati, R., and Smith, G.R., Eds., American Society for Microbiology, Washington, D.C., 1988, 231.

74. Smith, G.R., RecBCD enzyme, in *Nucleic Acids and Molecular Biology*, Vol. 4, Eckstein, F. and Lilley, D.M.J., Eds., Springer-Verlag, Berlin, 1990, 78.

75. Dixon, D.A., and Kowalczykowski, S.C., Homologous pairing *in vitro* stimulated by the recombination hotspot, Chi, *Cell*, 66, 361, 1991.

76. Dixon, D.A., and Kowalczykowski, S.C., The recombination hotspot, χ, is a regulatory sequence that acts by attenuating the nuclease activity of the *Escherichia coli* recBCD enzyme, *Cell*, 73, 87, 1993.

77. Dixon, D.A., Churchill, J.J., and Kowalczykowski, S.C., Reversible inactivation of the Escherichia coli recBCD enzyme by the recombination hotspot, χ, *in vitro*: evidence for functional inactivation or loss of the recD subunit, *Proc. Natl. Acad. Sci., U.S.A.*, 91, 2980, 1993.

78. Taylor, A.F., and Smith, G.R., Substrate specificity of the DNA unwinding activity of the recBC enzyme of *E. coli*, *J. Mol. Biol.*, 185, 431, 1985.

79. Roman, L.J., and Kowalczykowski, S.C., Characterization of the helicase activity of the *Escherichia coli* recBCD enzyme using a novel helicase assay, *Biochemistry*, 28, 2863, 1989.

80. Taylor, A., and Smith, G.R., Unwinding and rewinding of DNA by the recBC enzyme, *Cell*, 22, 447, 1980.

81. Roman, L.J., Eggleston, A.K., and Kowalczykowski, S.C., Processivity of the DNA helicase activity of *Escherichia coli* recBCD enzyme, *J. Biol. Chem.*, 267, 4207, 1992.

82. Telander-Muskavitch, K.M., and Linn, S., Electron microscopy of *E. coli* recBC enzyme reaction intermediates, in *Mechanistic Studies of DNA Replication and Genetic Recombination*, Vol. 19, Alberts, B., Ed., Academic Press, New York, 1980, 901.

83. Telander-Muskavitch, K., and Linn, S., RecBC-like enzymes: exonuclease V deoxyribonucleases, in *The Enzymes*, Boyer, P.D., Ed., Academic Press, New York, 1981, 233.

84. Telander-Muskavitch, K., and Linn, S., A unified mechanism for the nuclease and unwinding activities of the recBC enzyme of *E. coli*, *J. Biol. Chem.*, 257, 2641, 1982.

85. Roman, L.J., and Kowalczykowski, S.C., Characterization of the adenosine–triphosphatase activity of the *Escherichia coli* recBCD enzyme: relationship of ATP hydrolysis to the unwinding of duplex DNA, *Biochemistry*, 28, 2873, 1989.

86. Eggleston, A.K., O'Neill, T.E., Bradbury, E.M., and Kowalczykowski, S.C., Unwinding of nucleosomal DNA by the *Escherichia coli* recBCD helicase, submitted, 1993.

87. Nakayama, H., Nakayama, K., Nakayama, R., Irino, N., Nakayama, Y., and Hanawalt, P.C., Isolation and genetic characterization of a thymineless death-resistant mutant of *Escherichi coli* K12: identification of a new mutation (*recQ1*) that blocks the *recF* recombination pathway, *Mol. Gen. Genet.*, 195, 474, 1984.

88. Umezu, K., Nakayama, K., and Nakayama, H., *Escherichia coli* RecQ protein is a DNA helicase, *Proc. Natl. Acad. Sci. U.S.A.*, 87, 5363, 1990.

89. Umezu, K., and Nakayama, H., The recQ DNA helicase of *Escherichia coli*. Characterization of the helix-unwinding activity with emphasis on the effect of single-stranded DNA binding protein, *J. Mol. Biol.*, 230, 1145, 1993.

90. Tsaneva, I.R., Muller, B., and West, S.C., RuvA and RuvB proteins of *Escherichia coli* exhibit DNA helicase activity *in vitro*, *Proc. Natl. Acad. Sci. U.S.A.*, 90, 1315, 1993.

91. Tsaneva, I.R., Muller, B., and West, S.C., ATP-dependent branch migration of Holliday junctions promoted by the RuvA and RuvB proteins of *E. coli*, *Cell*, 69, 1171, 1992.

92. Tsaneva, I.R., Illing, G., Lloyd, R.G., and West, S.C., Purification and properties of the RuvA and RuvB proteins of *Escherichia coli*, *Mol. Gen. Genet.*, 235, 1, 1992.

93. Iwasaki, H., Takahagi, M., Nakata, A., and Shinagawa, H., *Escherichia coli* RuvA and RuvB proteins specifically interact with Holliday junctions and promote branch migration, *Genes Dev.*, 6, 2214, 1992.

94. Müller, B., Tsaneva, I.R., and West, S.C., Branch migration of Holliday junctions promoted by the *Escherichia coli* ruvA and ruvB proteins. I. Comparison of ruvAB- and ruvB-mediated reactions, *J. Biol. Chem.*, 268, 17179, 1993.

95. Müller, B., Tsaneva, I.R., and West, S.C., Branch migration of Holliday junctions promoted by the *Escherichia coli* ruvA and ruvB proteins. II. Interaction of ruvB with DNA, *J. Biol. Chem.*, 268, 17185, 1993.

96. Lloyd, R.G., and Sharples, G.J., Dissociation of synthetic Holliday junctions by *E. coli* recG protein, *EMBO J.*, 12, 17, 1993.

97. Lloyd, R.G., and Sharples, G.J., Processing of recombination intermediates by the recG and ruvAB proteins of *Escherichia coli*, *Nucleic Acids Res.*, 21, 1719, 1993.

98. Whitby, M.C., Ryder, L., and Lloyd, R.G., Reverse branch migration of Holliday junctions by RecG protein: a new mechanism for resolution of intermediates in recombination and DNA repair, *Cell*, 75, 341, 1993.

99. Joseph, J.W., and Kolodner, R., Exonuclease VIII of *Escherichia coli*. II. Mechanism of action, *J. Biol. Chem.*, 258, 10418, 1983.

100. Lovett, S.T., and Kolodner, R.D., Identification and purification of a single-stranded-DNA-specific exonuclease encoded by the *recJ* gene of *Escherichia coli*, *Proc. Natl. Acad. Sci. U.S.A.*, 86, 2627, 1989.

101. Lovett, S.T., DeLuca, C.L., and Kolodner, R.D., The genetic dependence of recombination in *recD* mutants of *Escherichia coli*, *Genetics*, 120, 37, 1988.

102. Iwasaki, H., Takahagi, M., Shiba, T., Nakata, A., and Shinagawa, H., *Escherichia coli* RuvC protein is an endonuclease that resolves the Holliday structure, *EMBO J.*, 10, 4381, 1991.

103. Dunderdale, H.J., Benson, F.E., Parsons, C.A., Sharples, G.J., Lloyd, R.G., and West, S.C., Formation and resolution of recombination intermediates by *E. coli* RecA and RuvC proteins, *Nature (London)*, 354, 506, 1991.

104. Bennett, R.J., Dunderdale, H.J., and West, S.C., Resolution of Holliday junctions by ruvC resolvase: cleavage specificity and DNA distortion, *Cell*, 74, 1021, 1993.

105. Connolly, B., and West, S.C., Genetic recombination in *Escherichia coli*: Holliday junctions made by RecA protein are resolved by fractionated cell-free extracts, *Proc. Natl. Acad. Sci. U.S.A.*, 87, 8476, 1990.

106. Sancar, A., Williams, K.R., Chase, J.W., and Rupp, W.D., Sequences of the *ssb* gene and protein, *Proc. Natl. Acad. Sci. U.S.A.*, 78, 4274, 1981.

107. Kowalczykowski, S.C., Bear, D.G., and von Hippel, P.H., Single-stranded DNA binding proteins, in *The Enzymes*, Boyer, P. D., Ed., Academic Press, New York, 1981, 373.

108. Muniyappa, K., Shaner, S.L., Tsang, S.S., and Radding, C.M., Mechanism of the concerted action of recA protein and helix destabilizing proteins in homologous recombination, *Proc. Natl. Acad. Sci. U.S.A.*, 81, 2757, 1984.

109. Kowalczykowski, S.C., and Krupp, R.A., Effects of the *Escherichia coli* SSB protein on the single-stranded DNA- dependent ATPase activity of *Escherichia coli* recA protein: evidence that SSB protein facilitates the binding of recA protein to regions of secondary structure within single-stranded DNA, *J. Mol. Biol.*, 193, 97, 1987.

110. Chow, S.A., Rao, B.J., and Radding, C.M., Reversibility of strand invasion promoted by recA protein and its inhibition by *Escherichia coli* single-stranded DNA-binding protein or phage T4 gene 32 protein, *J. Biol. Chem.*, 263, 200, 1988.

111. Lavery, P.E., and Kowalczykowski, S.C., A postsynaptic role for single-stranded DNA-binding protein in recA protein-promoted DNA strand exchange, *J. Biol. Chem.*, 267, 9315, 1992.

112. Smith, G.R., Hotspots of homologous recombination, *Experientia*, 50, 234, 1994.

113. Lam, S.T., Stahl, M.M., McMilin, K.D., and Stahl, F.W., *Rec*-mediated recombinational hot spot activity in bacteriophage lambda. II. A mutation which causes hot spot activity, *Genetics*, 77, 425, 1974.

114. Stahl, F.W., Crasemann, J.M., and Stahl, M.M., *Rec*-mediated recombinational hot spot activity in bacteriophage lambda. III. Chi mutations are site-mutations stimulating rec-mediated recombination, *J. Mol. Biol.*, 94, 203, 1975.

115. Dower, N.A., and Stahl, F.W., Chi activity during transduction-associated recombination, *Proc. Natl. Acad. Sci. U.S.A.*, 78, 7033, 1981.

116. Cheng, K.C., and Smith, G.R., Distribution of Chi-stimulated recombinational exchanges and heteroduplex endpoints in phage lambda, *Genetics*, 123, 5, 1989.

117. Smith, G.R., Comb, M., Schultz, D.W., Daniels, D.L., and Blattner, F.R., Nucleotide sequence of the Chi recombinational hotspot Chi+D in bacteriophage lambda, *J. Virol.*, 37, 336, 1981.

118. Ponticelli, A.S., Schultz, D.W., Taylor, A.F., and Smith, G.R., Chi-dependent DNA strand cleavage by recBC enzyme, *Cell*, 41, 145, 1985.

119. Taylor, A.F., Schultz, D.W., Ponticelli, A.S., and Smith, G.R., RecBC enzyme nicking at Chi sites during DNA unwinding: location and orientation-dependence of the cutting, *Cell*, 41, 153, 1985.

120. Roman, L.J., and Kowalczykowski, S.C., Formation of heteroduplex DNA promoted by the combined activities of *Escherichia coli* recA and recBCD proteins, *J. Biol. Chem.*, 264, 18340, 1989.

121. Roman, L.J., Dixon, D.A., and Kowalczykowski, S.C., RecBCD-dependent joint molecule formation promoted by the *Escherichia coli* recA and SSB proteins, *Proc. Natl. Acad. Sci. U.S.A.*, 88, 3367, 1991.

122. Wang, T.C., and Smith, K.C., The roles of RecBCD, Ssb and RecA proteins in the formation of heteroduplexes from linear-duplex DNA *in vitro*, *Mol. Gen. Genet.*, 216, 315, 1989.

123. Kowalczykowski, S.C., and Roman, L.J., Reconstitution of homologous pairing activity dependent upon the combined activities of purified *E. coli* recA, recBCD, and SSB proteins, in *Molecular Mechanisms in DNA Replication and Recombination*, Vol. 127, Richardson, C.C., and Lehman, I. R., Eds., Wiley-Liss, New York, 1990, 357.

124. Smith, G.R., Schultz, D.W., Taylor, A.F., and Triman, K., Chi sites, recBC enzyme, and generalized recombination, *Stadler Symp.*, 13, 25, 1981.

125. Thaler, D.S., Sampson, E., Siddiqi, I., Rosenberg, S.M., Stahl, F. W., and Stahl, M., A hypothesis: Chi-activation of recBCD enzyme involves removal of the recD subunit, in *Mechanisms and Consequences of DNA Damage Processing*, Friedberg, E., and Hanawalt, P., Eds., Alan R. Liss, Inc., New York, 1988, 413.

126. Stahl, F.W., Thomason, L.C., Siddiqi, I., and Stahl, M.M., Further tests of a recombination model in which chi removes the RecD subunit from the RecBCD enzyme of *Escherichia coli*, *Genetics*, 126, 519, 1990.

127. Korangy, F., and Julin, D.A., Kinetics and processivity of ATP hydrolysis and DNA unwinding by the recBC enzyme from *Escherichia coli*, *Biochemistry*, 32, 4873, 1993.

128. West, S.C., Cassuto, E., and Howard-Flanders, P., RecA protein promotes homologous-pairing and strand-exchange reactions between duplex DNA molecules, *Proc. Natl. Acad. Sci. U.S.A.*, 78, 2100, 1981.

129. West, S.C., Countryman, J.K., and Howard-Flanders, P., Enzymatic formation of biparental figure-eight molecules from plasmid DNA and their resolution in *E. coli*, *Cell*, 32, 817, 1983.

130. Connolly, B., Parsons, C.A., Benson, F.E., Dunderdale, H.J., Sharples, G.J., Lloyd, R.G., and West, S.C., Resolution of Holliday junctions *in vitro* requires the *Escherichia coli* ruvC gene product., *Proc. Natl. Acad. Sci. U.S.A.*, 88, 6063, 1991.

131. Hsieh, P., Camerini-Otero, C.S., and Camerini-Otero, R.D., Pairing of homologous DNA sequences by proteins: evidence for three-stranded DNA, *Genes Dev.*, 4, 1951, 1990.

132. Shibata, T., DasGupta, C., Cunningham, R.P., and Radding, C.M., Purified *E. coli* recA protein catalyzes homologous pairing of superhelical DNA and single-stranded fragments, *Proc. Natl. Acad. Sci. U.S.A.*, 76, 1638, 1979.

133. McEntee, K., Weinstock, G.M., and Lehman, I.R., Initiation of general recombination catalysed *in vitro* by the recA protein of *Escherichia coli*, *Proc. Natl. Acad. Sci. U.S.A.*, 76, 2615, 1979.

134. Radding, C.M., Beattie, K.L., Hoffman, W.K., and Wiegand, R.C., Uptake of homologous single-stranded fragments by superhelical DNA IV branch migration, *J. Mol. Biol.*, 116, 825, 1977.

135. Rigas, B., Welcher, A.A., Ward, D.C., and Weissman, S.M., Rapid plasmid library screening using RecA-coated biotinylated probes, *Proc. Natl. Acad. Sci. U.S.A.*, 83, 9591, 1986.

136. Konforti, B.B., and Davis, R.W., DNA substrate requirements for stable joint molecule formation by the RecA and single-stranded DNA-binding proteins of *Escherichia coli*, *J. Biol. Chem.*, 266, 10112, 1991.

137. Ferrin, L.J., and Camerini-Otero, R.D., Selective cleavage of human DNA: RecA-assisted restriction endonuclease (RARE) cleavage, *Science*, 254, 1494, 1991.

138. Adzuma, K., Stable synapsis of homologous DNA molecules mediated by the *Escherichia coli* RecA protein involves local exchange of DNA strands, *Genes Dev.*, 6, 1679, 1992.

139. Revet, B.M., Sena, E.P., and Zarling, D.A., Homologous DNA targeting with RecA protein-coated short DNA probes and electron microscope mapping on linear duplex molecules, *J. Mol. Biol.*, 232, 779, 1993.

140. Zarling, D.A., and Arboleda, M.J., RecA protein-dependent homologous DNA targeting in metabolically active human nuclei monitored by fluorescent *in situ* hybridization and laser microscopy, in *Fourth Int. Cent. Biotechnol. Symp. Gene Transfer Gene Ther.*, Stockholm, Sweden, 1993.

141. Olson, M., Hood, L., Cantor, C., and Botstein, D., New game plan for genome mapping, *Science*, 245, 1434, 1989.

142. Koob, M., Burkiewicz, A., Kur, J., and Szybalski, W., RecA-AC: single-site cleavage of plasmids and chromosomes at any predetermined restriction site, *Nucleic Acids Res.*, 20, 5831, 1992.

143. Golub, E.I., Ward, D.C., and Radding, C.M., Joints formed by RecA protein from oligonucleotides and duplex DNA block initiation and elongation of transcription, *Nucleic Acids Res.*, 20, 3121, 1992.

144. Golub, E.I., Radding, C.M., and Ward, D.C., Inhibition of RNA polymerase II transcription by oligonucleotide-RecA protein filaments targeted to promoter sequences, *Proc. Natl. Acad. Sci. U.S.A.*, 90, 7186, 1993.

145. Koller, T., DiCapua, E., and Stasiak, A., Complexes of recA protein with single-stranded DNA, in *Mechanisms of DNA Replication and Recombination*, Cozzarelli, N. R., Ed., Alan R Liss, New York, 1983, 723.

146. Flory, J., Tsang, S.S., and Muniyappa, K., Isolation and visualization of active presynaptic filaments of recA protein and single-stranded DNA, *Proc. Natl. Acad. Sci. U.S.A.*, 81, 7026, 1984.

147. Zarling, D.A., Homologous DNA targeting, In *Second Int. Conf. Gene Ther. Cancer*, San Diego, 1993.

148. Pinkel, D., Straume, T., and Gray, J., Cytogenetic analysis using quantitative, high-sensitivity, fluorescence hybridization, *Proc. Natl. Acad. Sci. U.S.A.*, 83, 2934, 1986.

149. Kido, M., Yoneda, Y., Nakanishi, M., Uchida, T., and Okada, Y., *Escherichia coli* RecA protein modified with a nuclear location signal binds to chromosomes in living mammalian cells, *Exp. Cell Res.*, 198, 107, 1992.

150. Kowalczykowski, S.C., *In vitro* reconstitution of homologous recombination reactions, *Experientia*, 50, 204, 1994.

151. Jackson, D.A., and Cook, P.R., A general method for preparing chromatin, *EMBO J.*, 4, 913, 1985.

152. Hall, S.D., and Kolodner, R.D., personal communications.

Chapter 8

Gene Targeting in Human Gene Therapy

Manuel A. Vega

Laboratory of Gene Therapy

Department of Biology

Universidad Nacional del Sur

San Juan, Argentina

Contents

0-8493-8950-X/95/$0.00+$.50

I. Introduction

In the preceding chapters, the use of gene targeting is described as a precise tool for inactivating or mutating selected genes in order to assess their function in the cell. In these cases, functional genes are inactivated through homologous recombination with a targeting DNA that carries a region of homology with the target locus and foreign non-homologous (interrupting) sequences.

The human genome shows a high degree of variability, and mutations are found all along it. Most mutations on the DNA are neutral in the sense that their occurrence does not interfere with the normal functioning of the cells. Some mutations, however, alter the functionality of the affected loci leading to the absence or dysfunction of the gene products involved and, eventually, to the pathogenesis of a (genetic) disease.

The feasibility of interrupting a gene inside a cell through its homology with a targeting DNA paves the way to the reverse process: namely, the correction of mutations already present on the target locus with a non-mutant targeting DNA. The phenomenon finds a potential application in the treatment of genetic diseases. Such an approach could be considered as the ideal form of the so-called gene therapy.

The concept of gene therapy has evolved over the last two decades.[1] The principle is simple: for a disease caused by mutations on a given gene, a genetic treatment can be envisaged consisting in the "replacement" of the mutant gene with an active copy of it. In this concept, replacement of the mutant gene has different meanings: replacement of (only) its activity or replacement of the gene itself. That difference defines two types of gene therapy: (1) gene complementation and (2) gene correction.

Gene complementation implies the introduction into the affected cells of an active copy (the "therapeutic gene") of the mutant gene. Precise positioning of the new DNA copy along the genome is, in principle, irrelevant. The endogenous gene remains mutant and untouched. The product of the therapeutic gene complements the gene function previously lacking, while the mutant protein continues to be produced from the endogenous locus. The therapeutic gene is introduced through genetically engineered replication-defective viral vectors of different types: retroviruses, adenoviruses, AA-viruses, and herpes viruses.[2-5]

By gene correction, on the contrary, we identify the process by which the introduced copy of a gene, or a part of it, replaces the mutant endogenous gene *in situ* at its precise genomic location.[6] The product of the introduced copy is produced in place of that of the mutant gene, which no longer exists.

From the point of view of function, we can define a third type of approach: gene addition therapy. In this case, active copies of a gene not naturally expressed nor even present in the genome are introduced in the recipient cells. As a result, a new genetic function is added, instead of a preexisting one being replaced or complemented. Gene addition therapy strategies are being developed for acquired diseases like AIDS and cancer.

On a functional basis, we can divide proteins in two groups: those functioning in *trans* and those functioning in *cis* with respect to tissue. A given protein functions in *cis* with respect to tissue when the direct effects of its activity are limited to the tissue where it is being expressed (for instance, cystic fibrosis transmembrane conductance regulator, dystrophin, β-globin, etc.). On the contrary, a protein functions in *trans* with respect to tissue when its activity can also directly affect other tissues, apart from the one where it is expressed (for example adenosine deaminase, α1-antitrypsin, factor IX, etc.).

Gene therapy of mutations on proteins with function in *cis* with respect to tissue, must be addressed to the tissue naturally expressing the mutant gene. On the contrary, gene therapy approaches to gene products with function in *trans* with respect to tissue can make use, in theory, of any tissue as a recipient for the expression of the therapeutic gene. We can define *cis*- and *trans*-strategies for gene therapy according to whether the tissue genetically manipulated is the one naturally expressing the mutant gene or not, respectively. Proteins with function in *cis* with respect to tissue admit only *cis*-strategies, while proteins with function in *trans* with respect to tissue admit either *cis*- or *trans*-strategies of gene therapy.

In this review, we will concentrate on the gene therapy approach that relies upon gene targeting events or gene correction therapy. As defined in the preceding chapters, gene targeting is the procedure of artificially addressing a specific genetic locus and of introducing changes into it. The only way a genetic locus can be specifically addressed and distinguished from the rest of the genome is based on its specific sequence. Once the introduced DNA has recognized the target locus and Holliday structures have been formed, a precise transfer of information can take place between the two participating molecules by homologous recombination upon successful resolution of the Holliday intermediates.[7] Although usually employed as equivalent to homologous recombination, gene targeting strictly means the event of addressing a locus, but not what happens thereafter. Reactions other than homologous recombination can follow a gene targeting event: (1) The targeting DNA can hybridize to the target locus giving rise to a triple-stranded DNA. The target locus can then react with chemical groups coupled to the targeting DNA.[8] So far, by this means, however, no specific changes in the sequence of the target locus have been obtained. (2) Alternatively, the targeting DNA can hybridize at the target locus by classic Watson-Crick pairing. If a mismatch between targeting DNA and target locus exists, a heteroduplex will be formed. Repair of the heteroduplexes may lead to specific changes in the target locus,[9] although it can also be affected by erratic mutations.

Although there are different ways of modifying a genetic locus by gene targeting, homologous recombination is the method that appears to have a major potential for application in gene therapy. In this chapter, we will refer to either gene targeting or homologous recombination as representing the same phenomenon.

II. Appealing features of gene targeting

Expression of the individual genes is highly regulated and relies on a precise program controlling developmental time, tissue distribution, and physiological triggers for expression. Regulation of gene expression is ensured by several DNA regions located either inside or surrounding the gene itself or far away from it. Most of the regulatory regions need to be precisely located relative to the regulated gene in order to be effective.

By a gene complementation therapy approach, introduction of a therapeutic gene at any site on the genome would need to provide the introduced gene with every regulatory DNA region necessary to ensure the level and distribution of expression required.[10] Key DNA regions must then be identified, cloned, and characterized. On the contrary, by gene targeting (as it involves the correction of the mutation *in situ*), the corrected gene product is expressed under its natural regulation, as tight as it might be. In this case, the respective regulatory DNA regions do not even need to be identified.

From the scope of gene correction therapy, a mutant locus can be considered just as a genetically inert sequence of DNA. Gene correction is essentially a structural manipulation with functional consequences. The fact that homologous recombination takes place only on a homology basis makes gene correction ideally applicable, whatever the function of the mutant DNA may be: a coding region, a regulatory region, etc. Gene targeting is, in principle, applicable to any form of disease, either recessive or dominant. Gene complementation, on the contrary, is basically a functional manipulation. With the simple procedure of transferring a complementing copy of the mutant gene by gene complementation, only recessive diseases can be approached. Dominant pathologies, either genetic or acquired, need more elaborated strategies of either gene complementation or addition. For instance, specific ribozymes are being tested for their potential to degrade specific RNAs inside transfected cells.[11] If such an approach is to be applied to dominant genetic disorders, a ribozyme specific for the mutant, but not for the normal, mRNA allele has to be designed. A misrecognition of the normal mRNA allele by the ribozyme would not eliminate the mutant protein, and in addition, it would degrade the normal message.

Upon gene correction, the natural physical and functional relationships of the mutant locus with the rest of the genome are conserved. Therefore, the stability of the locus, either stability of expression or genetic stability, remains unchanged. The corrected gene will be expressed naturally as long as the genetic program allows it to happen and will be structurally stable, at least as stable as the respective normal locus. For the current gene complementation strategies, a stable expression of the therapeutic gene is difficult to obtain. Expression of the therapeutic gene is usually controlled by foreign bacterial or viral regulatory elements, and it is usually affected by the presence of viral vector sequences that can even block its expression.[12,13] Structural stability is, in addition, not a common outcome in gene complementation experiments, especially when retrovirus vectors are used for gene transfer.[14]

In theory, there are no size limits for the targeting DNA molecule. On the contrary, the size of a DNA fragment that can be transferred with the current gene complementation technology is restricted to what can be efficiently cloned and transduced using the viral vectors employed. The approach becomes limited to diseases involving genes with active forms sized in the range accomodable inside the vectors used.

Finally, it has recently been shown that upon homologous recombination, no changes in the sequence of the target locus occur other than the preplanned alteration itself.[16] Thus, in addition to its high degree of specificity, the fidelity of the gene targeting reaction makes gene correction a procedure adequately safe for therapeutic purposes. Besides that, no other particular safety considerations concern gene correction therapy. On the other hand, as long as gene complementation relies on the use of viruses, the issue of safety has to be stressed. The potential risks associated with the use of viruses, either for the patient, the gene therapists, or the general public remain latent at this time. In addition, the integration of the therapeutic gene at random sites along the genome in every new transduced cell introduces the risk of oncogenesis by insertional mutagenesis. (With the best current technology, however, that possibility is theoretically associated with a very low probability and experimentally not observed yet.)[15]

Taken together, the preceding points concerning stability of expression, genetic stability, safety, control of time and tissue, distribution of expression, and fidelity make gene targeting by homologous recombination an ideal way to test gene therapy for human genetic diseases.

III. Drawbacks of gene targeting

In the preceding section, gene targeting has been described as the ideal way to accomplish the genetic treatment of human genetic diseases. However, there are several reasons that make gene targeting a very difficult task in the context of gene therapy.

The technology and the principles applied to gene correction by homologous recombination are the same that have been described in the foregoing chapters for gene disruptions. None of the methods developed so far allows for performing gene targeting *in vivo*. Therefore, mutant cells have to be explanted and cultured or maintained *in vitro* if gene targeting is to be performed. For a number of relevant diseases, however, (for instance, those involving most nervous tissue) either (1) the cell type implicated could hardly be manipulated *ex vivo*, or (2) the integrity of the tissue involved cannot be disrupted without affecting its function, or (3) the highly organized tissular structures would not regenerate upon reimplantation of corrected cells. For these reasons, genetic manipulation will preferably have to be performed *in vivo*. No attempts to modify any tissue *in vivo* by gene targeting have been reported yet. Viral vectors for mediation of *in vivo* delivery and activity of targeting DNAs might be developed in the future.[17,18] However, the use of viruses as vectors for the targeting DNA would expose the field of gene correction to safety concerns like those related to virus-mediated gene complementation therapy. Alternatively, empty viral capsides might be employed for transferring *in vivo* the targeting DNA into the target cell nucleus with high efficiency and avoiding the inclusion of viral sequences.[19]

Following gene correction, the corrected gene is indistinguishable from its normal allele and its expression relies on the same factors and events that control the expression of the normal allele. Correction of a given gene in cells not expressing it under physiological conditions will not lead to the expression of the active protein, even when the mutation might have been reverted at the DNA level. Therefore, gene correction has to be carried out in cells where the target gene is naturally expressed under physiological conditions. Thus, gene correction is adequate only for *cis*-strategies of gene therapy. For gene correction, but not for gene complementation, *trans*-strategies are not feasible even when proteins with function in *trans* with respect to tissue are implicated.

If other genomic regions apart from the target locus itself share high homology with the targeting DNA, they might compete with the real target locus. Pseudogenes, members of gene families, the allelic loci of the target gene, and shared domains between the target locus and other genes are examples of such potentially interfering DNA regions.

Although the gene targeting reaction is highly accurate and the cells undergoing it do not suffer other changes in their genomes, other cells in the population of transfected cells usually do. The targeting DNA, as any transfecting DNA, can be inserted randomly at any position along the genome by so-called illegitimate recombination. Illegitimate recombination does not lead to correction of the target gene, and cells undergoing it will continue expressing a mutant phenotype. In mammalian cells, gene targeting occurs about a thousand times less frequently than does illegitimate recombination. Illegitimate recombination should somehow be blocked or the cells that have suffered it should be eliminated. Otherwise, the major argument for gene targeting (namely, that it occurs *in situ* on the target locus itself) is diluted by a degenerate reaction taking place a thousand times more frequently. Attempts aimed at decreasing the ratio of absolute frequencies between random insertion and homologous recombination have been reported.[20,21]

The targeting DNA has to span over the mutant region itself, which might not include nor surround the coding region of the gene affected by the mutation. For many diseases, the mutation will probably be different from patient to patient. The location of the mutation (although not necessarily the nature of the mutation itself) for every new patient must first be determined, and the respective normal (non-mutant) genomic DNA fragment must be made available as targeting DNA. On the contrary, for a gene complementation approach, a common cDNA copy of the therapeutic gene suffices as the only specific reagent in the transfer reaction for every putative recipient of the therapy.

The limitations of gene targeting described above, although certainly relevant, do not actually impact the current state of the art in the field. The major difficulty for gene targeting on its way towards gene therapy is the efficiency of the targeting reaction itself. In mammalian, including human, cells, the absolute frequency of gene targeting is around 1 targeted cell out of 10^6 transfected cells. Different methods (double selection, promoter-less, etc.; see preceding chapters) have been developed that decrease the relative frequency of illegitimate recombination and result in the apparent increase of gene targeting. However, as far as gene therapy is concerned, the real need is for methods that might increase the absolute and not (only) the relative frequency of targeting events. Nuclear microinjection has been reported to allow for an up to 10^4-fold increase in the absolute frequency of targeting events. Significant increases in the absolute frequency of targeting, although still irrelevant in a gene therapy context, can also be obtained with targeting DNAs isologous to their respective target loci, by increasing the length of homology between targeting DNA and target locus, and by using linear targeting DNA molecules (see Chapter 1 and Chapter 2). Finally, there is evidence that genes are targeted more efficiently by insertion vectors than by replacement vectors.[22]

Replacement vectors exchange the DNA fragment surrounded by the homologous arms between the targeting DNA and the target locus(see Chapter 1). Replacement vectors in which the homologous arms surround just the corresponding normal (wild type) sequence are, in theory, the ideal vectors for gene correction purposes.

Insertion vectors, on the other hand, are not appropriate because they lead to the insertion of the whole vector, as well as to a partial duplication of the homologous arms at the target locus. The two-step gene targeting reaction known as hit-run or, in general, in-out (see preceding chapters) makes use of an insertion vector in order to precisely modify a few bases on the target locus. This in-out procedure is the only one so far tested that allows for a preplanned alteration of a target sequence at the base pair level. Nevertheless, its application in gene correction therapy is not straightforward, mainly because two cycles of *in vitro* culture, selection and amplification of the target cells, should be completed. Moreover, it is not suited for strategies involving *in vivo* gene correction.

With the current technology, as mentioned, gene correction can be made only on explanted cells. Because of the extremely low efficiency of gene targeting, most of the cells in a transfected population remain unchanged. The bulk of explanted cells cannot, therefore, be directly reimplanted following transfection, as it is done after viral mediated gene transfer. Instead, cells of the particular target cell type should be first highly enriched and, following transfection, specifically recovered.

Recovery of those infrequent cells that underwent gene correction can be performed either by (1) a selection procedure that favored survival of targeted cells or (2) the specific detection of the corrected cells by a non-destructive screening procedure. Both alternatives rely on a cell phenotype arising after the correction event and require that the target locus (either the target gene itself or an introduced marker) be expressed in the population of transfected cells. In both cases, extensive *in vitro* culture would be required for multiplication either of the primitive explants or of the selectively recovered targeted cells. The procedure becomes, then, less attractive because of the risks of cell transformation associated with *in vitro* culture itself[23] and because of the loss of the polyclonality of the initial cell population (polyclonality is desirable, for instance, when lymphoid tissue is involved).

In practice, neither direct selection nor screening of corrected cells are easy to document. Screening might be based on labeling cells with antibodies against epitopes expressed on the corrected, but not on the mutant, gene product of interest. That requires, however, that the pathogenic mutation directly or indirectly changes an extracellular protein domain, which is obviously not always the case. When secretory proteins are involved, screening for the corrected cells might also be based on the detection of a local concentration of the secreted substrate around the corrected cells on cell monolayers. Airway epithelium cells, for instance, express CFTR, the protein implicated in cystic fibrosis. Cells expressing active CFTR (CFTR$^+$ cells) secrete Cl^- in response to cAMP stimulators, while cells from cystic fibrosis patients (CFTR$^-$) do not. Higher Cl^- potentials can be measured on the surface of CFTR$^+$ cells compared to CFTR$^-$ cells when cell monolayers are scanned with a Cl^--electrode (Vega, Goossens and Besmond, unpublished[43]). The same principle could be applied to screening procedures for some other diseases.

Direct selection methods, in which selection is based on the activity of the corrected gene, allow, at once, for the recovery of the targeting reaction product itself. That is the case of, for instance, severe combined immunodeficiency (SCID(ADA$^-$), cystic fibrosis, and the Lesch-Nyhan syndrome. ADA$^+$ (adenosine deaminase positive) cells can be recovered in medium containing adenosine analogs, while ADA$^-$ patient cells die.[25] CFTR$^+$ airway epithelium cells, but not CFTR$^-$ cells from cystic fibrosis patients, selectively resist a treatment in a medium containing epinephrine.[24] Finally, HPRT$^+$ cells can be selectively grown in an appropriate culture medium,[26] while Lesch-Nyhan ($HPRT^-$) cells die. Direct selection methods are restricted to a few diseases in which expression of the corrected allele confers to the cell a selective advantage.

Some indirect methods of selection, like in-out procedures, have been already described. In another approach, a selective marker is introduced at the time the mutant gene is corrected. As a result, targeted (corrected) cells become selectable through the activity of the marker.[27] This procedure requires that the presence and activity of the selectable marker not alter the expression level and pattern, the genetic stability, nor the regulation of the target locus.

The potential application of gene correction therapy is currently restricted to cells that can be explanted, cultured, and maintained *in vitro* in order to complete the necessary steps of transfection, selection/screening, and amplification. Tissue-specific stem cells are the ideal cell target because they have the potential to originate or reconstitute cell lineages, thus giving rise to a lasting therapeutic effect upon reimplantation. For most tissues, however, stem cells have not been identified nor characterized yet. With our current knowledge and technology, *in vitro* culture and reimplantation of stem cell-containing tissue explants can be performed only for a very few tissues like bone marrow and muscle.

Diseases affecting muscular tissue, like Duchenne muscular dystrophy (DMD), are interesting targets for gene therapy and much emphasis is being placed on them. However, myoblasts (muscular stem cells) have to be reimplanted into every muscular piece of tissue. Disorders involving the diverse hemopoietic tissue are less complicated. Handling of bone marrow stem cells becomes less and less difficult[28] and reimplantation can be accomplished simply through the blood stream. Diseases affecting tissues derived from the bone marrow are particularly good candidates for gene targeting-mediated gene therapy.

A recent approach makes use of complexes between targeting DNA and bacterial recombinogenic proteins to increase the absolute frequency of targeting events. RecA protein plays a key role in homologous recombination in *E. coli* (see preceding chapters). In addition, cloning and characterization of recA-like proteins of human origin have been started.[29] Attempts have been made to increase the frequency of $CFTR^+$ cells after transfection of $CFTR^-$ human tracheal epithelium cells with CFTR-cDNA coated with *E. coli* recA protein.[30] In order to preserve the activity of the complexes, direct nuclear microinjection or encapsidation into viroid particles are the transfection procedures of choice. If the DNA/protein complexes turn out to be effective for increasing gene targeting frequency, their use will greatly facilitate the identification and recovery of the corrected cells. These DNA-RecA protein pioneer complexes will probably have to be supplemented with additional proteins also participating in the recombination process. It can be expected that in the future more complex particles will be designed that are able to carry out not only the targeting reaction itself, but also cell intake and transport of the DNA/recombinogenic protein complex into the nucleus. Such particles, which might make use of real viral elements, would allow for *in vivo* strategies of gene targeting.

IV. Germ cells engineering

Some genetic disorders show clinical symptoms clearly concentrated on certain tissues. For instance, Duchenne muscular dystrophy affects muscle cells and thalassemias affects blood cells. In such cases, expression of the genes involved is restricted to a limited set of cell lineages: the dystrophin gene in muscle cells and the globin genes in red blood cells. Mutations on these genes are actually neutral for any other cell types and will not cause a mutant phenotype on them. For these types of disorders, either gene complementation or gene correction approaches could be applied exclusively on the affected tissue(s). Unaffected tissues do not need to be genetically manipulated.

Other disorders arise by mutations on genes expressed in several, or even most, tissues. For example, SCID(ADA$^-$) is caused by mutations on the ADA gene and cystic fibrosis by mutations on the CF gene. The ADA gene is expressed in most tissues. Although lymphocytes are the cells the most sensitive to ADA-deficiency, other cell types, like fibroblasts, are also sensitive to mutations on the ADA gene.[31,32] As ADA is a protein that functions in *trans* with respect to tissue, a gene therapy *trans*-strategy, that is, expression of ADA in any cell, would eliminate at once any mutant phenotype on the affected tissues.

The situation is different for proteins with functions in *cis* with respect to tissue and for the diseases they cause, like CFTR and cystic fibrosis.[33] Clinical symptoms of cystic fibrosis involve several tissues: respiratory, intestinal, gonadal, and hepatic. The most prominent phenotype is the respiratory affection caused by the absence of active CFTR on the airway epithelium. As CFTR functions in *cis* with respect to tissue, only gene therapy *cis*-strategies can be envisaged. Reversion of the respiratory mutant phenotype by gene therapy will not reverse the affection of, for instance, the intestinal epithelium. Then, following a successful airway gene therapy, we would be faced with a different pathology affecting principally the gastrointestinal tissue, and so on.

For those types of diseases, a definitive cure should reach every one of the different tissues affected. That can only be accomplished by genetic treatment of the cell(s) from which the different affected tissues stem: that is, early embryo somatic or of germ cells. Thus, either germ line or early embryo somatic cell gene therapy is the future challenging therapeutic answer to the disorders produced by proteins with function in *cis* with respect to tissue.

Somatic cells used in gene therapy strategies are either terminally differentiated cells (i.e. tracheal epithelium, lymphocytes, neurons), cells that will undergo differentiation along a defined terminal pathway (i.e. hemopoietic stem cells, myoblasts, hepatocytes), or cancer cells. Aberrant genetic products that might arise from gene complementation treatments, like insertional mutagenesis, would spread to a more or less limited population of cells, and would never be passed to the offspring. However, the situation is not the same when germ cells or early embryo somatic cells are manipulated. Germ cells give rise, at every offspring generation, to a complete new somatic body and to the continuity of the germ cell line. The potential of the germ cells is total. Early embryo somatic cells are pluripotent and will, thus, also give rise to many different cell lineages and types. The overall structure and function of every region of the human genome are not yet known. Thus, spurious alterations of the genome on those very primitive cells might have deleterious and unforeseeable consequences for many different differentiated cell lineages. Because the genetic modification will need to be properly expressed at the right times and places and will be passed to every cell of the body and, more importantly, also to the offspring, the manipulation should be as clean and safe as possible. Gene targeting, in spite of its drawbacks, is the only foreseeable way to perform such a manipulation of the genome.[34]

When viral vectors are used in a somatic cell gene therapy strategy, the simultaneous transfer of the therapeutic gene to a bulk of cells is performed regardless of the individual destination of each particular recipient cell. Success is achieved not (only) by taking care of individual target cells, but by increasing the number of cells that are transduced at once. Such a procedure works in the context of somatic gene therapy. When germ line gene therapy is concerned and human embryos are involved, however, the situation is essentially different. There is no need for massive treatments for gene transfer. On the contrary, germ line manipulation will have to be performed on an extremely reduced number of cells, and each of them becomes important and unique. Their fate will have to be surviving and pursuing development. It is essential to the concept of germ cell gene therapy to take care of and to preserve every particular target zygote or embryo. Otherwise, easier procedures than gene therapy on the germ line, like gamete or zygote selection, should be envisaged.

Finally, the fact that genetic manipulation of either the zygote (germ line) or of early embryo somatic cells can be carried out *ex vivo* and that such cells are suitable for direct nuclear microinjection (with either naked DNA or DNA/protein complexes) brings germ line gene therapy closer into the area technically feasible by gene targeting.

V. Animal models

Testing in animals is required before gene correction therapy strategies can become clinically applicable. Concerning gene correction, particular considerations can be made about the parameters to be tested in animals and the requirements for an animal to be a reliable model.

The first step includes the evaluation of the molecular genetics of the process, like the gene targeting reaction itself, the parameters affecting it, and the features of the targeting DNA, etc. This step also allows for optimizing conditions for selection/screening of corrected cells.

In the gene complementation therapy field, the same viral vectors that will finally be used on the human patient cells can be tested in animal models.

Moreover, current viral vectors make use of regulatory signals linked to the therapeutic gene that are not specific for human cells. Therefore, expression of the therapeutic gene in the cells of the animal model will closely reflect the expression it will show in the human target cells.

For gene correction, the situation is more complex. The homologous recombination reaction depends on endogenous cellular activities varying with the cell differentiation stage, the cell type, and the transformation state of the cell.[35,36] Moreover, the chromosome and the genome localization of the target mutation may affect its capability to undergo recombination and/or the accessibility of the recombinogenic proteins. Gene targeting relies on sequence homology between targeting DNA and target locus. Because the sequence of the target locus will diverge from species to species, especially when no coding regions are involved, the targeting DNA cannot be the same molecular species when the target locus is human as opposed to animal. Moreover, human regulatory regions (like promoters, enhancers, splice signals, etc.) included in the targeting DNA, might not be equally active in the animal cells.

For these reasons animal models are not adequate for the first step. Instead, it needs to be performed in human tissue/type-specific cells, either the same or close to those to be finally used in the clinical protocol.

The second step involves *in vivo* testing. That is assessment of (1) the therapeutic efficacy of the corrected cells as well as of the particular gene therapy protocol; (2) the efficiency of recovery, survival, etc. of the corrected cells upon reimplantation, and (3) the deleterious or risky effects that reimplantation itself or the activity of the corrected cells might cause in the recipient.

For the evaluation of those parameters, human cells are no longer necessary and animal models become valuable as long as they reproduce the physiopathology of the human disease. Such parameters of the therapy can be evaluated by reimplantation of the appropriate normal animal cells in the mutant animal model to assess what the effects of introducing human normal (corrected) cells into a human (mutant) patient would be. No genetically manipulated cells are even necessary at this step.

An interesting mouse model currently employed for the detection and quantitation of mutagens is the λ/β-gal carrying mouse, like MutamouseTM and others.[37] These mice carry many copies of λ-phage DNA, each carrying a copy of the *E. coli* β-galactosidase wild-type gene. When DNA from Mutamouse tissue is isolated, packaged into λ capsides, and plated on bacteria, λ/β-gal DNA will give rise to blue plaques unless mutations have occurred on the β-gal gene itself. The number of white plaques is a measure of the frequency of mutational events along the mouse genome. Using mutant β-gal DNA as targeting DNA, the frequency of white plaques will measure the occurrence of homologous recombination between the normal target locus and the mutant targeting DNA. Instead, mice carrying a mutant β-gal gene (in place of the wild type) targeted with a wild-type β-gal targeting DNA would greatly decrease the background of false positive plaques. Mutamouse-like mice are, in theory, useful models to assess direct *in vivo* gene targeting into different mouse tissues, including gonads.

VI. Some genes of interest

So far, a few genes of medical relevance have been targeted and their mutations corrected by homologous recombination.

The HPRT (hypoxanthine guanine phosphoribosyl transferase) gene was the first gene corrected by gene targeting.[26] The mutant HPRT locus was targeted with a piece of DNA carrying a noncomplementary mutation on the HPRT fragment. $HPRT^+$ (corrected) cells were selectively recovered in medium containing 6-thioguanine. The product of the targeting reaction, the $HPRT^+$ cells, can be directly recovered; no screening procedures nor selectable markers in the targeting DNA are necessary. The HPRT gene is expressed in most tissues, but mainly in some regions of the brain. Mutations on the HPRT gene cause the pathology characterized by degeneration of the central nervous system known as Lesch-Nyhan Syndrome. At birth, irreversible affection of the brain is already present. Neurons are the cells primarily affected by the disease. Successful treatment of Lesch-Nyhan Syndrome by gene therapy should be performed on nervous tissue at the embryo stage. Thus, proceeding from the successful *in vitro* studies showing that correction of the HPRT gene is feasible towards the development of a clinical protocol becomes very hampered.

Another gene that has been corrected is the β-globin gene. Hybrid mouse-human cells carrying a mutant β^S-globin gene were transfected with a targeting DNA containing a piece of the wild-type β^A-globin sequence linked to the neo^r gene.[27] The targeting DNA was made in such a way that, upon homologous recombination with the target β^S-globin locus, the β^S mutation was corrected into the β^A normal allele, and at the same time the neo gene became inserted on one side of the β-globin sequence. Corrected cells were selected by their resistance to G418 conferred by the selectable marker. So far, no method has been developed for the direct selection of cells expressing the normal β-globin allele. Mutations on the globin genes are associated to thalassemias, and in particular, the mutation β^S is the cause of sickle-cell anemia, the most frequent genetic disorder among black people. The target tissues for any gene therapy approach against thalassemias or sickle-cell anemia are bone marrow stem cells. Gene complementation approaches are hampered because a precise co-regulation of both α and β genes would be required, and current vectors do not allow for such a level of precision. By contrast, gene correction would keep the endogenous regulation on the corrected globin genes. Targeting bone marrow stem cells with a targeting DNA carrying a selectable marker expressed in the stem cells upon recombination at the β-globin target locus is the most promising gene correction therapy approach.

Cystic fibrosis, the most frequent genetic disorder among caucasians, is of great interest as well. It is caused by mutations on the CF gene, whose product (CFTR) is a transmembrane cAMP-dependent Cl^- channel. The most conspicuous and lethal clinical symptoms occur in the respiratory tract, where the airway epithelium is involved. Several different cell types are found in the epithelium and presumably also stem cells, although they have not yet been characterized. Recently, it has been shown that retroviral vector-mediated gene transfer into explanted tracheal epithelial rat cells is feasible and, more importantly, that the genetically manipulated cells can be reimplanted back into recipient rats, giving rise to a mosaic of endogenous and transplanted epithelial cells.[38] The experiment shows that *ex vivo* therapy for cystic fibrosis, although technically complicated, should not be excluded. By a mechanism not elucidated, application of epinephrine on human tracheal epithelial cells cultured *in vitro* selectively kills $CFTR^-$ (cystic fibrosis) cells, while equivalent $CFTR^+$ (normal) cells survive the treatment. This fact is at the basis of a selection method for the direct recovery of corrected ($CFTR^+$) cells arising by gene targeting.[24] Correction of the most frequent mutation (ΔF508) on the CFTR by gene targeting on human tracheal epithelial cells *in vitro* has been reported.[30] In summary, because CFTR mutant airway epithelium cells can be corrected *in vitro*, because $CFTR^+$ cells can be positively selected for, and because transplantation of airway epithelium seems to be feasible, gene correction therapy for cystic fibrosis might become feasible by *ex vivo* manipulation of airway epithelial cells. Much work must still be performed, however, before such a possibility becomes clinically significant.

Another genetic disorder we will consider here is SCID(ADA^-),[31] a very rare disease caused by mutations on the ADA gene. It is clinically characterized by a lethal combined immunodeficiency, presumably caused by the accumulation of ADA substrates. Several treatments have been assayed for SCID(ADA^-), including ADA^+ erythrocyte transfusion, purified ADA injection, and gene complementation therapy on T and bone marrow cells. All these treatment strategies take advantage of the fact that ADA functions in *trans* with respect to tissue. Gene correction therapy approaches to SCID(ADA^-) would be aimed at the correction of the ADA gene in bone marrow stem cells.

Several features make ADA deficiency a prime candidate for clinical trials of gene correction therapy: (1) correction of either of the two alleles would reverse the mutant phenotype of the cells, as SCID(ADA^-) is a recessive disorder; (2) a reliable direct selection procedure for ADA^+ cells is available; (3) the target cells are the hemopoietic stem cells (as mentioned, bone marrow tissue is well suited for *ex vivo* gene therapy and the putative stem cells can be highly purified); and (4) some evidence supports the expectation that ADA^+ (lymphoid) cells have a selective advantage over ADA^- cells when introduced into ADA-deficient patients.[39] Thus, a few corrected (ADA^+) stem cells would presumably survive and be able to repopulate the hemopoietic tissue.

SCID(ADA$^-$), together with globin disorders, will probably be one of the first diseases to be treated by gene correction therapy. Globin disorders share the profitable features of SCID(ADA$^-$), except for the selective advantages of corrected cells either *in vitro* or (more importantly) *in vivo*. However, a good targeting efficiency on stem cells coupled to the insertion of a selective marker should overcome that limitation.

VII. Gene addition

The possibility of targeting a DNA fragment into a preselected specific locus has other important implications in human gene therapy, apart from the treatment of genetic diseases by gene correction. As mentioned, gene addition therapy is the transfer of genes coding for products (either protein or RNA) not present as such in the recipient cell.

Gene addition therapy strategies against AIDS and cancer are being developed.[40,41] In spite of its therapeutic effectiveness, the practice of randomly inserting vectors, like retroviruses, might lead to undesired insertional activations/inactivations. The AAV (Adenovirus-Associated Virus) insertion of the provirus, on the other hand, occurs at a unique and specific site on the human genome, in chromosome 19.[42] Interestingly enough, insertion of the provirus in chromosome 19 does not seem to produce any unwanted insertional activation or inactivation. For that reason, AAVs, in spite of their particular drawbacks, have rapidly gained interest as vectors for gene therapy.

Looking for site-specific insertion in order to avoid undesired insertional mutagenesis, gene targeting emerges again as the ideal way to perform gene addition. A neutral locus, namely, a locus that might function as target for heterologous genes, could be targeted with a DNA in the form of an insertion vector. The homologous arms of the vector would ensure specific targeting, and the heterologous part, the therapeutic gene, would perform the expected therapeutic function.

VIII. Gene correction and genetic polymorphisms

Polymorphisms, or silent mutations, are accumulated all along the genome, principally in regions not under selective pressure, like introns and intergenic fragments. It has recently been suggested that use of targeting DNAs isogenic to the target loci leads to an important increase in the targeting efficiency, presumably caused by the perfect sequence matching between targeting DNA and target locus.

Human cells are diploid, and most alleles, because they come from different progenitors, are polymorphic each other. Rigorously speaking, a DNA is only isogenic (identical) to a given allele: the isogenic DNA to a mutant allele from a given patient is the DNA from the same mutant allele of that patient. Thus, isogenic targeting DNA cannot be used to correct a mutation (because it is mutant in itself), unless the mutation is first eliminated by *in vitro* engineering.

In spite of its effect on the targeting efficiency, the presence of polymorphisms between targeting DNA and target locus has another potential application. Polymorphisms generate the so-called RFLP (Restriction Fragment Length Polymorphims) that can be detected on the DNA by Southern blot analysis. By homologous recombination between a targeting DNA having a given RFLP pattern and a locus carrying a different RFLP pattern, a new (recombined) pattern of RFLPs will be generated. The targeted locus, and hence the corrected cells, might, thus, be identified by its own RFLP pattern. As an alternative procedure for identifying corrected cells, the only limiting condition is that the RFLP patterns of the targeting DNA and the target locus have to be previously known and must be different from each other.

IX. Final remarks

The most appealing feature of gene targeting is that the regulation and the pattern of expression of the corrected gene are conserved. For gene complementation approaches, reaching that goal is certainly a very difficult task.

In addition, the interest in gene correction therapy is sustained on a perfectionist principle that intends to avoid alterations of the genome and of the target locus with foreign DNA sequences. However, improvement of gene complementation tools might reach a state where the unexpected genetic changes they may cause will become insignificant compared with the high degree of polymorphisms and plasticity of the genome, and that the risks of using viral vectors will not be higher than those caused by spontaneous mutations nor by the extensive *in vitro* culture necessary to recover gene-targeted cells.

The key issue to be addressed for gene targeting to become an alternative for gene therapists is the efficiency of the homologous recombination event. As long as the targeting reaction relies on the endogenous enzymatic machinery of the target cell, it might be that there is no way to increase such efficiency up to useful values. On the other hand, any attempts to use exogenous recombinases will introduce new variables and risks that might render the gene targeting in human gene therapy impracticable.

References

1. Anderson, F.W., Prospects for human gene therapy, *Science*, 26, 401, 1984.

2. Donehower, L.A., Research and potential clinical applications of retroviral vectors, *Prog. Med. Virol.*, 34, 1, 1987.

3. Stratford-Perricaudet, L., Perricaudet, M., Gene transfer into animals: the promise of adenovirus, in *Human Gene Transfer*, Vol. 219, Cohen-Haguenauer, O., Boiron, M., Eds., Coll.INSERM, 1991, 51.

4. Muzyczka, N., Use of adeno-associated virus as a general transduction vector for mammalian cells, *Curr. Top. Microbiol. Immunol.*, 158, 97, 1992.

5. Fink, D.J., Sternberg, L.R., Weber, P.C., Mata, M., Goins, W.F., Gloriosos, J.C., *In vivo* expression of β-galactosidase in hippocampal neurons by HSV-mediated gene transfer, *Hum. Gene Ther.*, 3, 11, 1992.

6. Vega, M.A., Prospects for homologous recombination in human gene therapy, *Hum. Genet.*, 87, 245, 1991.

7. Muller, B., Jones, C., Kemper, B., West, S.C., Enzymatic formation and resolution of Holliday junctions *in vitro*, *Cell*, 60, 329, 1990.

8. Riordan, L.M., Martin, J.C., Oligonucleotide-based therapeutics, *Nature*, 350, 442, 1991.

9. Weiss, U., Wilson, J.H., Heteroduplex-induced mutagenesis in mammalian cells, *Nucleic Acids Res.*, 16, 2313, 1988.

10. Dzierzak, E.A., Papayannopoulou, T., Mulligan, R.C., Lineage-specific expression of a human β-globin gene in murine bone marrow transplant recipients reconstituted with retrovirus-transduced stem cells, *Nature*, 331, 35, 1988.

11. Weerasighe, M., Liem, S.E., Asad, S., Read, S.E., Joshi, S., Resistance to human immunodeficiency virus type 1 (HIV-1) infection in human CD4+ lymphocyte-derived cell lines conferred by using retroviral vectors expressing an HIV-1 RNA-specific ribozyme, *J. Virol.*, 65, 5531, 1991.

12. Berkner, K.L., Expression of heterologous sequences in adenoviral vectors, *Curr. Top. Microbiol. Immunol.*, 158, 39, 1992.

13. Soriano, P., Friedrich, G., Lawinger, P., Promoter interactions in retrovirus vectors introduced into fibroblasts and embryonic stem cells, *J. Virol.*, 65, 2314, 1991.

14. Emerman, M., Panganiban, A.T., Temin, H.M., Insertion of genes into retrovirus genomes and of retrovirus DNA into cell genomes, *Prog. Med. Virol.*, 32, 174, 1985.

15. Anderson, W.F., What about those monkeys that got T-cell lymphoma?, *Hum. Gene Ther.*, 4, 1, 1993.

16. Zheng, H., Hasty, P., Brenneman, M.A., Grompe, M., Gibbs, R.A., Wilson, J.H., Bradley, A., Fidelity of targeted recombination in human fibroblasts and murine embryonic stem cells, *Proc. Natl. Acad. Sci. U.S.A.*, 88, 8067, 1991.

17. Boggs, S.S., Targeted gene modification for gene therapy of stem cells, *Int. J. Cell Cloning*, 8, 80, 1990.

18. Salmons, B., Gunzburg, W.H., Targeting of retroviral vectors for gene therapy, *Hum. Gene Ther.*, 4, 129, 1993.

19. Gao, L., Wagner, E., Cotten, M., Agarwal, S., Harris, M., Romer, M., Miller, L., Hu, P.C., Curiel, D., Direct *in vivo* gene transfer to airway epithelium employing adenovirus-polylysine-DNA complexes, *Hum. Gene Ther.*, 4, 17, 1993.

20. Chang, X.-B., Wilson, J.H., Modification of DNA ends can decrease end joining relative to homologous recombination in mammalian cells, *Proc. Natl. Acad. Sci. U.S.A.*, 84, 4959, 1987.

21. Waldman, B.C., Waldman, A.S., Illegitimate and homologous recombination in mammalian cells: differential sensitivity to an inhibitor of poly(ADP-ribosylation), *Nucleic Acids Res.*, 18, 5981, 1990.

22. Hasty, P., Rivera-Pérez, J., Chang, C., Bradley, A., Target frequency and integration pattern for insertion and replacement vectors in embryonic stem cells, *Mol. Cell. Biol.*, 11, 4509, 1991.

23. Hayflick, L., Moorhead, P.S., The serial cultivation of human diploid cell strains, *Exp. Cell Res.*, 25, 585, 1961.

24. Vega, M.A., Goossens, M., Besmond, C., A powerful method for *in vitro* selection of normal versus cystic fibrosis human airway epithelial cells, *Gene Ther.*, 1, 1, 1993.

25. Kantoff, P.W., Kohn, D.B., Mitsuya, H., Armentano, D., Sieberg, M., Zwiebel, J.A., Eglitis, M.A., McLachlin, J.R., Wiginton, D.A., Hutton, J.J., Horowitz, S.D., Gilboa, E., Blaese, R.M., Anderson, W.F., Correction of adenosine deaminase deficiency in cultured human T and B cells by retrovirus-mediated gene transfer, *Proc. Natl. Acad. Sci. U.S.A.*, 83, 6563, 1986.

26. Thompson, S., Clarke, A.R., Pow, A.M., Hooper, M.L., Melton, D.W., Germ line transmission and expression of a corrected HPRT gene produced by gene targeting in embryonic stem cells, *Cell*, 56, 313, 1989.

27. Shesely, E.G., Kim, H.-S., Shehee, W.R., Papayannopoulou, T., Smithies, O., Popovich, B.W., Correction of a human β^S-globin gene by gene targeting, *Proc. Natl. Acad. Sci. U.S.A.*, 88, 4294, 1991.

28. Srour, E.F., Brandt, J.E., Briddell, R.A., Leemhuis, T., van Besien, K., Hoffman, R., Human CD34+ HLA-DR-bone marrow cells contain progenitor cells capable of self-renewal, multilineage differentiation, and long-term *in vitro* hematopoiesis, *Blood Cells*, 17, 287, 1991.

29. Yoshimura, Y., Morita, A., Yamamoto, A., Matsushiro, A., Cloning and sequence of the human recA-like gene cDNA, *Nucleic Acids Res.*, 21, 1665, 1993.

30. Gruenert, D.C., personal communication.

31. Martin, D.W., Biochemistry of diseases of immunodevelopment, *Ann. Rev. Biochem.*, 50, 845, 1981.

32. Palmer, T.D., Hock, R.A., Osborne, W.R.A., Miller, D., Efficient retrovirus-mediated transfer and expression of a human adenosine deaminase gene in diploid skin fibroblasts from an adenosine deaminase-deficient human, *Proc. Natl. Acad. Sci. U.S.A.*, 84, 1055, 1987.

33. Boat, T.F., Welsh, M.J., Beaudet, A.L., Cystic fibrosis, in *The Metabolic Basis of Inherited Disease*, Scriver, C.L., Beaudet, A.L., Sly, W.S., Valle, D., Eds., McGraw-Hill, New York, 1989, 2649.

34. Neel, J.V., Germ-line gene therapy: another view, *Hum. Gene Ther.*, 4, 127, 1993.

35. Buerstedde, J.-M., Takeda, S., Increased ratio of targeted to random integration after transfection of chicken B cell lines, *Cell*, 67, 179, 1991.

36. Finn, G.K., Kurz, B.K., Cheng, R.Z., Shmookler Reis, R.J., Homologous plasmid recombination is elevated in immortaly transformed cells, *Mol. Cell. Biol.*, 9, 4009, 1989.

37. Hazleton Research Products, Inc., P.O. Box 7200, Denver, PA 17517

38. Engelhardt, J.F., Allen, E.D., Wilson, J.M., Reconstitution of tracheal grafts with a genetically modified epithelium, *Proc. Natl. Acad. Sci. U.S.A.*, 88, 11192, 1991.

39. Lenarsky, C., Parkman, R., Bone marrow transplantation for the treatment of immune deficiency states, *Bone Marrow Transpl.*, 6, 361, 1990.

40. Sullenger, B.A., Gallardo, H.F., Ungers, G.E., Gilboa, E., Overexpression of TAR sequences renders cells resistant to human immunodeficiency virus replication, *Cell*, 63, 601, 1990.

41. Rosenberg, S.A., Anderson, W.F., Blaese, M.R., Immunization of cancer patients using autolocancer cells modified by insertion of the gene for tumor necrosis factor. Clinical protocol, *Hum. Gene Ther.*, 3, 57, 1992.

42. Samulski, R.J., Zhu, X., Xiao, X., Brook, J.D., Housman, D.E., Epstein, N., Hunter, L.A., Targeted integration of adeno-associated virus (AAV) into human chromosome 19, *EMBO J.*, 10, 3941, 1991.

43. Vega, M.A., Goossens, M., Besmond, C., 1993, unpublished.

INDEX

N

O

P

R

S

T

U

V

W

X

Y

Z